SOUTHERN
URIE & MAUNSELL
2-CYLINDER 4-6-0s

David Maidment

PEN & SWORD
TRANSPORT

First published in Great Britain in 2016 by
Pen & Sword Transport
An imprint of Pen & Sword Books Ltd
47 Church Street
Barnsley
South Yorkshire
S70 2AS

ISBN 978 1 47385 253 2

All royalties from this book will be donated to the Railway Children charity [reg. no. 1058991] [**www.railwaychildren.org.uk**]

Typeset in Palatino by Pen & Sword Books Ltd
Printed and bound in China by Imago Publishing Limited

Pen & Sword Books Ltd incorporates the imprints of Pen & Sword Archaeology, Atlas, Aviation, Battleground, Discovery, Family History, History, Maritime, Military, Naval, Politics, Railways, Select, Social History, Transport, True Crime, and Claymore Press, Frontline Books, Leo Cooper, Praetorian Press, Remember When, Seaforth Publishing and Wharncliffe.

For a complete list of Pen and Sword titles please contact
Pen and Sword Books Limited
47 Church Street, Barnsley, South Yorkshire, S70 2AS, England
E-mail: enquiries@pen-and-sword.co.uk
Website: www.pen-and-sword.co.uk

Previous Publications:

Novels (Religious historical fiction)
The Child Madonna, Melrose Books, 2009
The Missing Madonna, PublishNation, 2012
The Madonna and her Sons, PublishNation, 2015

Novels (Railway fiction)
Lives on the Line, Max Books, 2013

Non-fiction (Railways)
The Toss of a Coin, PublishNation, 2014
A Privileged Journey, Pen & Sword, 2015
An Indian Summer of Steam, Pen & Sword, 2015
Great Western Eight-Coupled Heavy Freight Locomotives, Pen & Sword, 2015

Non-fiction (Street Children)
The Other Railway Children, PublishNation, 2012
Nobody Ever Listened To Me, PublishNation, 2012

Cover photo: King Arthur 30802 *Sir Durnore* at Stewarts Lane depot before leaving for Victoria and a train to the Kent Coast. This engine was one of the six-wheel tender Arthurs built in 1926 for the Brighton line, and was subsequently allocated to the Southern Railway's Eastern Section until the Kent Coast electrification in 1959. It was then provided with a 5,000 gallon bogie tender from a withdrawn locomotive and spent the last couple of years of its life working from Waterloo to Basingstoke, Salisbury and Bournemouth, 30 March 1959. (R.C.Riley)

ACKNOWLEDGEMENTS

I have drawn on the knowledge and experiences of many previous authors, whose books and articles are listed in the bibliography. I have added my own experience of these engines in the last decade of their existence, when I lived on the Western Section of British Railways' Southern Region and commuted daily by steam train between Woking and Waterloo through 1957 to 1961, for which in particular I owe a debt of gratitude to Surrey County Council who funded my season ticket as part of my annual grant when I was a student at University College London. And I include short reminiscences by a former Bricklayers Arms fireman, David Solly, who had experience of King Arthurs on the Southern Region's Eastern Section, and my friend and colleague Colin Boocock, who was trained and employed at Eastleigh Works over fifty years ago.

In addition to my own photographs mostly taken in the period of my commuting to college, I have relied on the photographs of several other expert photographers and collections, in particular Colin Boocock, John Hodge, P.H. Groom, and the collections of the Manchester Locomotive Society and photo archivist, Paul Shackcloth; of John Scott-Morgan and of Mike Bentley; R.C. Riley's black and white photos, now with the Transport Treasury; Colin and James Garrett's collection of photos by Rev A.W. Mace; H.C. Casserley's photos in the collection of his son, Richard; and the colour photos of R.C. Riley in Rodney Lissenden's safe keeping and Paul Chancellor's ColourRail collection, all of whom have allowed me to use their pictures free of charge or at a reduced fee as all the royalties from this book are being donated to the Railway Children charity which I founded in 1995, and which is a charity supported by both the professional and heritage railway industries. I thank them and also acknowledge the help of the editors and staff of Pen and Sword and John Scott-Morgan in particular. Lastly, once again I am indebted to my daughter, Helen Breeze, for assisting me with the scanning and presentation of many of the photographs.

CONTENTS

Chapter 1

OVERVIEW

This book is about the design, construction and operation of Robert Urie's and Richard Maunsell's 2-cylinder 4-6-0s of the H15, N15, S15 family, together with Maunsell's rebuilding of the London, Brighton and South Coast Railway (LB&SCR) Baltic tanks, which became class N15X. However, the story needs to be seen in the context of the legacy in the

Drummond four-cylinder G14 4-6-0 No.456 shortly after construction, c 1909. This engine was withdrawn in 1925 and parts used to construct Eastleigh Arthur 456 *Sir Galahad*, although the main component retained for the new engine would be the 'watercart' tender. (J.M. Bentley Collection)

locomotive department that Urie inherited from his predecessor, Dugald Drummond and the later developments by Bulleid and Riddles.

Drummond had died in November 1912 following a serious accident at work and Urie, in effect his assistant as Works Manager at Eastleigh, took over the reins of the management of the London & South Western's locomotive fleet at a difficult and critical time. Although Drummond had equipped the LSWR with a fleet of excellent 4-4-0s, his attempts at providing satisfactory 4-cylinder 6ft

0in wheeled 4-6-0s for the increasing passenger traffic – especially the heavily graded line west of Salisbury – had led to failure until the T14s were constructed. However, even with modifications by Maunsell, the latter never reached the heights of other contemporary 4-6-0s, being heavy on water and coal and prone to run hot. The E14 and F13s of 1904-7 and the smaller G and P14s of 1908-10, with their mixture of Stephenson slide valves for the inside cylinders and Walschaerts gear for the outside, were sluggish, with high fuel consumption, and hardly reached the level of performance of the 4-4-0s.

Robert Urie was appointed Chief Mechanical Engineer in January 1913 and immediately set about addressing some of the problems he found. He favoured robust simple designs and his H15 of 1914 was a 2-cylinder 4-6-0 with 6ft driving wheels and outside Walschaerts valve gear, with high running plates to enable easy access for maintenance.

He followed the success of this design in the latter stages of the First World War by planning for a new family of passenger and freight locomotives as soon as the Eastleigh Plant was released from manufacturing war equipment. Ten

Urie H15 4-6-0 No.485 at its home depot, Nine Elms, c1932. 485 was the last of the 1914 Urie series, built at Eastleigh in June of that year and was one of the engines with a Schmidt superheater, replaced by a Maunsell in 1928. It was equipped with straight smoke deflectors in August 1930 and was the first of the class to be withdrawn after collision damage, in April 1955, after a life of over forty years and mileage run of 1.3 million.
(J.M. Bentley Collection)

passenger N15s, a 4-6-0 with 6ft 7in coupled wheels capable of dealing with the increasing loads that were prevalent at the end of the war, were completed between August 1918 and November 1919. He completed the family of new simple designs with a freight 4-6-0 with 5ft 7in coupled wheels in 1920, the S15.

Urie retired in 1922 – he had been already fifty-eight when he was appointed Chief – and Richard Maunsell, the previous Chief Mechanical Engineer of the South Eastern & Chatham Railway

(SE&CR), took over, based at their workshops at Ashford. Maunsell and his team, picked exclusively from his Ashford staff, were well acquainted with the principles of locomotive boiler and valve

gear design that Churchward had introduced at Swindon, and had applied this advanced thinking in the design of the 'N' class Moguls and the River 2-6-4 tanks. Maunsell discovered quickly that

Urie N15 30748 *Vivien*. Built August 1922, withdrawn from Basingstoke shed in September 1957 and cut up at Eastleigh in October of that year. The engine is fitted with electric lighting, a consequence of its conversion to oil firing between 1946 and 1948. It is seen here at Eastleigh in the last year of its life, c1956
(Manchester Locomotive Society Collection [MLS])

30453 *King Arthur*, allegedly rebuilt from parts of Drummond's G14 453 by Richard Maunsell in 1925, seen here at Nine Elms before leaving to haul a Waterloo – Basingstoke semi-fast service. It is still equipped with its original 'watercart' tender from the G14, so the photo dates from around 1955, after which 30453 was attached to a bogie tender from a withdrawn locomotive.
(P.H. Groom)

despite the huge improvements Urie had brought to the 4-6-0 fleet of the LSWR, the performance of some of the new engines fell short of expectations, not all was yet as satisfactory as it should have been and the newly amalgamated Southern Railway found itself heavily criticised in the press. In particular, the N15 class frequently suffered from poor steaming and loss of time on important expresses that were in the public eye.

In the aftermath of the war, there had been limitations on what Urie could achieve at Eastleigh, but now Maunsell began to address some of the problems. He set about tests with the N15s and applied the valve setting principles he and his key

staff had learned from Churchward and produced a modified N15 design, starting with the theoretical rebuilding of the inadequate G and P14s, though it is probable that little of these engines was utilised in the construction of the Eastleigh Arthurs, 448–457. Urie had rebuilt the Drummond F13s to his H15 design, and Maunsell rebuilt the sole E14, 335, to the H15 model also. He then built some more H15s, 473-478 and 521-524, to the Urie basic design, but with his own tapered boilers.

With the immediate success of the Eastleigh Arthurs, the SR Board approved the construction of a fleet of similar locomotives from the North British Locomotive Company

in Glasgow, and these engines, known to all as the 'Scotchmen', together with the Eastleigh-built engines, became the backbone of the Southern Railway locomotive fleet between the two world wars. They immediately took over the running of the heavy boat trains between Victoria, Folkestone Harbour and Dover, and powered the main expresses from Waterloo to Bournemouth, Salisbury and Exeter. The Southern Board had appointed John Elliot as Publicity Officer to counter some of the adverse press and it was his suggestion that led to the naming of the N15s after King Arthur and the Knights of the Round Table.

The Southern Railway Board

then set its priorities on the electrification of the Brighton line as well as its extensive suburban service, so Maunsell was inhibited in his expenditure on steam power. His attempt at a more powerful 4-cylinder passenger engine was limited to just sixteen examples, and was initially not successful enough to displace the N15s. The King Arthurs therefore continued to hold sway until Maunsell built his very successful Schools 4-4-0s for the Hastings line and then, in the later 1930s after the Portsmouth line electrification, had sufficient

to replace the N15s on the main Bournemouth expresses. The N15s dominated the Exeter route and the South Eastern section boat trains until the Bulleid Pacifics arrived in significant numbers.

In the meantime, Maunsell had applied the same principles of design to the Urie S15s and produced his own design in 1927, locomotives that, although intended as freight engines, rapidly assumed the 'mixed traffic' role. This was partly because there was a shortage of modern steam power as the available budget money was

spent primarily on electrification throughout the 1930s.

Maunsell had built earlier some N15s with six-wheel tenders instead of the bogie design for the former LB&SC lines and upon electrification of the Brighton route in 1933, these N15s were released for traffic mainly on the former SE&CR routes, with some of the Scotchmen bogie tender engines coming to augment those on the Western Section.

Another consequence of the electrification of the Brighton line was the redundancy of the

Maunsell S15 30830 at Feltham near the end of its life in July 1964, before being sold to Woodham Brothers scrapyard in Barry and subsequently retained for preservation. Its frames and wheels currently reside on the North Yorkshire Moors Railway workshops at Grosmont, awaiting restoration.
(MLS Collection)

Preserved King Arthur 30777 *Sir Lamiel* in the BR livery it wore during the last four years of its British Railways career, at Loughborough station during the Great Central Railway's Autumn Steam Gala, 2013. (Author)

LB&SCR's fine Baltic tanks, 2327-2333, and Maunsell took the opportunity to convert them to 4-6-0s, outwardly similar in many ways to the N15s, although they never quite achieved their promise in the rebuilt form. However, they performed some useful work on the Bournemouth road on summer Saturday holiday traffic and with semi-fast services to Basingstoke, Salisbury and Southampton.

The consequence of Maunsell's locomotive policies enabled the Southern Railway to accelerate its key services in the 1930s and overcome much of the criticism that the press had levelled at them. For example, the Southern Railway restored the pre-war two-hour schedule from Waterloo to Bournemouth and these trains were dominated by Bournemouth and Nine Elms King Arthurs until

Schools became available from 1937-8. However, the Second World War produced new demands. Despite the influx of Bulleid's Pacifics, both the Merchant Navies and the lighter West Countries, traffic after the war built up to such levels, especially in the summer holiday season, that the Maunsell N15s continued to play an extensive role.

With the nationalisation of the railways in 1948, it was some time before the new standard classes were built in sufficient numbers to provide any relief to the Southern Region steam power position. The need for the new BR standard locomotives on the former Great Eastern and various LMS lines took priority, and the fact that so many of the additional power requirements on the Southern were seasonal meant that the Maunsell classes held their own throughout the 1950s. Indeed, the Urie N15s lasted until the mid-1950s and the H15s to the end of the decade. The Maunsell passenger classes – N15s, Schools and Lord Nelsons – all went by the end of 1962, because by that time many seasonal services were being cut back and two series of BR Standard 5s, subsequently taking on the names of the Urie King Arthurs, presided over many of the former duties of the Maunsell passenger engines. Indeed, these engines, 73080-9 and 73110-9, owed much to the concept of the 2-cylinder mixed traffic 4-6-0 with Walschaerts valve gear, first developed by Urie in 1914.

The Urie and Maunsell S15s lasted longer as no BR Standard freight locomotives were allocated to the Southern Region and some of the Maunsell S15s lasted until 1966, almost to the end of SR steam use. A few could still be found occasionally on semi-fast commuter services in 1964 and 1965.

The Urie and Maunsell N15s, H15s and S15s were simple but strong robust machines that did their work in a reliable fashion with acceptable maintenance costs, though rarely reaching the occasional brilliance of some of the express power on the other railways, or even some of the exceptional performances of the Bulleid Pacifics. Bulleid himself seemed to decry the N15s as 'old fashioned' workhorses, too simple and straightforward to interest him, but there is no doubt that they performed invaluable work on top link expresses for nearly twenty years and continued to serve in a supportive role throughout the 1950s. A survey of the Western Section of the Southern Region on a summer Saturday in July 1956 found that Urie and Maunsell King Arthurs were more prevalent in operations that day than any other class of locomotive by a considerable margin.

Just one N15 has been preserved, the justly famous 777 *Sir Lamiel* which is credited with the one exceptional performance that matched the high speed exploits of some other railways. A number of Urie and Maunsell S15s have been preserved as a result of their greater longevity and subsequent sale to the Woodham Bros. scrapyard at Barry. No H15s or N15Xs survive.

This book will look in much greater detail at the Urie and Maunsell designs, the reasons for their creation and their performance in traffic, including some personal experiences of the author of all three classes of both designers during his travels in the 1950s and 1960s, and will incorporate the account of a Bricklayers Arms fireman, David Solly, of a journey on one of the depot's King Arthurs on a continental perishables ferrywagon train. Many previous books about these locomotives only deal with one of the classes or concentrated on a particular aspect, for example outward appearance and liveries for modellers, or their performance on the road, and in this book I attempt to bring such information together to give a comprehensive review of these classes and the Maunsell N15X rebuilds, coloured by my own personal experience and an update on the status of the preserved engines as they stand at the publication of this book.

Chapter 2

DRUMMOND'S LEGACY

Drummond 4-cylinder G14 prototype 453 in works grey livery for its official photograph, 1908.
(MLS Collection)

Drummond reigned as Chief Mechanical Engineer of the London & South Western Railway (L&SWR) from 1897 to 1912, having previously held that position on the Caledonian Railway, where he had designed a number of successful 4-4-0s. He repeated his success by building the T9 and L12 class 4-4-0s for the L&SWR passenger services, but his mind was turning to more ambitious outcomes. As well as planning and implementing the major move of the company's main workshop from Nine Elms to Eastleigh, he began to map out a number of theoretical improvements to tackle the increasing passenger traffic of the Edwardian era.

One of his experiments was an 1897-built 4-2-2-0, No.720, with 6ft 7in coupled wheels, bearing a similarity with Webb's London & North Western (L&NWR) compound locomotives, but with four high pressure cylinders rather than compounding. This was followed in 1901 by five rather larger engines, as E10 class, Nos. 369-373. They were unreliable and were said to only appear in traffic at peak periods.

In 1905 Drummond produced his first 4-6-0s for the L&SWR, the F13 class, 330–334. Two inside cylinders drove the first pair of 6ft driving wheels and the two outside drove the second pair – the driving wheels being coupled, unlike his earlier experiments. Another, 335, with minor differences, appeared in 1907, classified as an E14, and

in 1908 five more class G14s, 453-457, and in 1910, a final batch of P14s, 448-452. Although designed primarily to haul heavier loads over the steeply graded Salisbury–Exeter main line, they were sluggish and suffered from poor steaming, cracked frames, frequent hot axle boxes and were heavy on coal and water compared with the 4-4-0s which therefore still dominated the L&SWR's express passenger traffic. The 1911-built T14 4-cylinder 4-6-0s were better, but still were little improvement over the 4-4-0s, until rebuilt by Maunsell in 1930.

Basically these engines were under-boilered and complex and maintenance was difficult because of the inaccessibility of much of the running gear. The Outdoor Locomotive Department disliked the complications of the cross-water tubes in the firebox, steam feed pumps, exhaust steam heating of tender water and 'steam-dryers' in the smokebox, many of these parts being particularly inaccessible. Although Drummond had been a brilliant engineer, he became dogmatic and reluctant to hear and react to criticisms of his larger locomotives, despite the feedback from his operating staff. The 4-6-0s had 6ft driving wheels and a working pressure of 175psi. The running gear was massive and caused problems to the shed staff.

P14 No.449 as constructed at Eastleigh Works in 1910, before release for the traffic department, in its initial L&SWR livery.
(J.M. Bentley Collection)

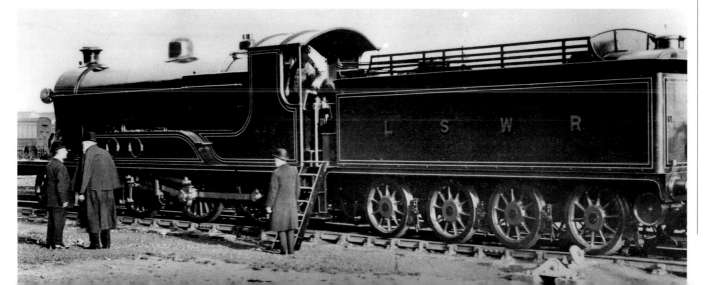

A group of L&SWR officers, including Dugald Drummond himself in the cape, inspect the newly-constructed P14 No.449 at Eastleigh, 1910.
(J.M. Bentley Collection/Real Photographs)

1905-built Drummond F13 No.331 at Eastleigh
Works together with a repaired L&SWR 0-6-0, No.341.
(J.M. Bentley Collection/Real Photographs)

P14 452 at Raynes Park with a Waterloo – Salisbury – Exeter express c1912.
(MLS Collection/G. Coltas)

When given special attention the engines could perform adequately, but in general service they were huge disappointments. Coal consumption was in excess of 50lb per mile and they were soon relegated to fast goods services between London and Southampton Docks. The E14, No.335, in particular, was notorious for its coal consumption and the authorities found that they had to roster two firemen to this engine.

The G and P14s were smaller, with Walschaerts valve gear to the outside cylinders and Stephenson slide valve gear to the inside cylinders. The G14s appeared first in the spring of 1908 and were immediately allocated to Salisbury to run the expresses west of the city. They were better than the earlier 4-6-0s but seldom improved on the performance of the 4-4-0s, even when skilfully fired and driven. They were successful enough, however, for Drummond to get authority to build five more with a slightly longer wheelbase and piston valves to the cylinders. Tests of a G14 and P14 between Salisbury and Exeter, against an L12 4-4-0, were favourable to the L12.

Finally Drummond produced a 4-cylinder 4-6-0 with 6ft 7in

E14 4-6-0 No.335 posed in works grey livery, 1907.
(J.M. Bentley Collection/Real Photographs)

coupled wheels in an endeavour to construct an engine that was more free-running downhill. The T14 was adequate for the Waterloo–Bournemouth and Salisbury services of the day, although the weaknesses of the axleboxes and high coal consumption had not yet been overcome. This design also had a higher boiler pressure – 200psi – than the earlier 4-6-0s.

Cecil J. Allen enjoyed a number of footplate runs between Salisbury and Exeter in 1912 and he commented in his regular 'Locomotive Running Past and Present' article in the November 1950 *Trains Illustrated* that he noted little with these big engines (the P14s and G14s) that a T9 couldn't have done almost as well. He quoted a couple of runs with 455 on the 9am Waterloo–Exeter and the fast 7.20am Exeter–Waterloo with 451 when time was barely maintained with loads of less than 300 tons.

Drummond had also produced a further 4-4-0, the D15, which was an appreciable improvement on the 4-6-0s and could manage a heavier load than the excellent T9s. With the fastest trains east of Salisbury and on the Bournemouth route in the hands of the D15s, L12s and T9s, Robert Urie, when he was appointed as the L&SWR's Chief Mechanical Engineer, had inherited an excellent new Works at Eastleigh, capable 4-4-0s, but indifferent 4-6-0s that struggled to maintain time on the fastest services and which had already been relegated to freight work for much of their existence. The newer fleet also had unacceptable running and maintenance costs.

F13 334 passes Eastleigh with a heavy freight for Southampton Docks. Eastleigh Carriage Works is in the background. Photo taken c1912. (J.M. Bentley Collection)

A Drummond F13, No.330, in traffic at Eastleigh shed alongside a K10 4-4-0, 8 May 1920. (J.M. Bentley Collection/H.C. Casserley)

Superheated 4-cylinder T14 No.462 before rebuilding by Maunsell, in the mid-1920s.
(MLS Collection)

Drummond T14 No. 458, as rebuilt by Maunsell, with the distinctive 'Paddlebox' splashers removed. 458 was a casualty of a direct bomb hit on Nine Elms depot during the Second World War.
(J.M. Bentley Collection/ Photomatic)

T14 No.462 passes Esher with a Salisbury–Waterloo semi-fast train, July 1931.
(J.M. Bentley Collection)

T14 444 on a long Up Goods train between Farnborough and Woking, c1945.
(John Scott-Morgan Collection)

T14 444 poses with another of its class and West Country 34023 *Blackmore Vale* at Nine Elms c1949. (MLS Collection)

One of the two surviving T14s to be painted in BR black livery with its new number, 30461, at Nine Elms c1950. This is the form in which the author observed these engines working the 3.54pm Waterloo- Basingstoke in 1949/50 when he came out of school at Surbiton.
(J.M. Bentley Collection)

Chapter 3

ROBERT WALLACE URIE

Robert Urie was born at Ardeer in Ayrshire on 22 October 1854 and was educated at Ardeer and Glasgow High School. He became an apprentice with the firm Gauldie, Marshall & Co., then at Dubs & Co., and finally at William King & Co., a six-year stint from 1869 to 1875. He attended the Glasgow Mechanics Institution, and was awarded the first prize for mechanical drawing for the session 1871–1872. During the years from 1875 to 1890 he was engaged as a draughtsman by A. Barclay & Sons, and Grant Ritchie and Company of Kilmarnock, Hawthorn Davey & Company of Leeds, and the Clyde Locomotive Company before joining the Caledonian Railway, as a draughtsman under the Chief, Dugald Drummond. In 1890 he became the chief draughtsman at St Rollox Works and in 1896 he became the Works Manager.

Urie continued as Works Manager at St Rollox when Drummond left the Company to spend a brief period in Australia, prior to his appointment as Locomotive Engineer for the London and South Western Railway in 1897. The same year, Urie joined Drummond in a move to the L&SWR, becoming Works Manager at Nine Elms, before moving to Eastleigh in 1909 to manage the newly opened Works there. Clearly, despite Drummond's reputation as an autocrat and difficult boss, Urie had a good relationship with him and respected the Chief Mechanical Engineer. Drummond must also have had a regard for Urie's abilities, bringing him south as part of his immediate top team. During this period Urie lived with his family in Larkhall Rise, South London. In 1898, he applied for and was granted Membership of the Institution of Mechanical Engineers.

Drummond apparently got his feet wet one autumn day and demanded a mustard bath to restore circulation. The initial water was not hot enough, and the next lot was too hot and caused scalding, and the damaged skin turned septic. Drummond continued at work but the leg had become gangrenous and required amputation. He died on 7 November 1912. In January 1913, Urie was appointed as the Chief Mechanical Engineer of the L&SWR, even though he was already fifty-eight years of age. His loyalty to Drummond must have been put to the test during the final years of Drummond's tenure, when he would have been aware of the shortcomings of the latter's 4-6-0s, but as soon as he was in the chair at Eastleigh, he nailed his colours to the mast with a series of designs that were almost the opposite extreme of Drummond's final efforts – designs that elevated simplicity, robustness and ease of maintenance to the fore.

He had been an efficient and effective Works Manager at both St Rollox and Eastleigh and soon after his new L&SWR appointment he found himself putting Eastleigh Works on a war footing, organising it to manufacture munitions. He also served on the committee of locomotive engineers brought together to design some standard locomotives for the railways of the UK and service overseas on the front line.

The number of anecdotes about the character of the larger-than-life Dugald Drummond are legion, but the stories about Urie are far fewer. However, this does not mean he was a weaker character under the influence and patronage of the Old Man (as Drummond was known to his subordinates). He would not have flourished under Drummond if he had been unable to stand up to his chief. He was said to have had a phenomenal memory, which his subordinates may have found inconvenient at times, and he had a reputation of being able to see right through his staff to their discomfort, especially if they were guilty of some error or misdemeanour. Apparently it was his custom to

The late Mr. R.W. Urie, Locomotive Engineer, London & South Western Railway, 1912-23, scanned from the *Railway Gazette* of January 15 1937 and attributed to Elliott & Fry. The photograph was in the magazine's obituary as the death of Robert Urie occurred earlier that month, and is printed here by courtesy of Mr Urie's grandson, also Robert Urie.

walk regularly through the Works, speaking to no-one and seeming to observe nothing. Then, at the end of his tour, various managers and foremen would be summoned to receive the full force of his wrath over some inefficiency or poor practice that he had observed.

Urie was a most able engineer. A man of strong personality, he could command in an effortless manner. He wasted few words, his utterances were apparently short and to the point. Stern he might have been, in the management culture of the period, but he had the reputation also of being a very sincere man and very dignified. After the war, he set about a long-term plan to put the locomotive department of the L&SWR on a more secure footing. He had already set out his principles of simplicity and robustness in the design and construction of the H15 2-cylinder 4-6-0 in 1914 and designs for a passenger and freight 2-cylinder 4-6-0 were pursued during the period when Eastleigh's capacity was filled with locomotive repairs and munitions work. In January 1919 he presented to the Company's Board a five-year plan, indicating the need for:

45 N15 4-6-0s

25 H15 4-6-0s (including the rebuilding of five Drummond F13s)

40 S15 4-6-0s

The superheating of Drummond T9s, S11s, L12s and '700' class 0-6-0s

8 G16 heavy shunting engines

10 H16 goods tank engines

12 six-coupled shunting engines

Although he was to retire at the 'Grouping' in 1923, much of his five-year plan was fulfilled, with Maunsell completing the order of S15s, and some of the H15s.

Urie was a great proponent of superheating, and as well as testing different types of superheater on his first design (the H15), he set about the equipping of many older engines with significant improvement in efficiency, as shown above in his five-year plan.

During the First World War and its immediate aftermath and despite all the difficulties of the wartime priorities and lack of resources for locomotive building, Urie managed to construct seventy-five new and reliable engines that, with the Drummond 4-4-0s, were the bedrock of the L&SWR's motive power fleet. Indeed, his concept of a 2-cylinder 4-6-0 with outside Walschaerts valve gear for general traffic duties of all types became the standard mixed traffic locomotive for later company designs and ultimately the fleet of BR Standard 4-6-0 locomotives, some of which perpetuated symbolically the names with which his own passenger 4-6-0s were later bestowed.

At the time of the Grouping Urie was already 68 years old and perhaps it was inevitable that he would face retirement in view of the need to find one new Chief Mechanical Engineer to co-ordinate the activities of the locomotive policy of the three companies that amalgamated to become the Southern Railway.

During his time at Eastleigh, he and his family lived in Hill Lane, Southampton. Three of his sons – James, David and John – joined the railway industry and all spent some time at Eastleigh. James was Assistant Works Manager, and subsequently went to the railways of Chile. David was in the Drawing Office and subsequently, after a period on the Midland & Great Western of Ireland, returned to the London Midland and Scottish Railway in the Locomotive Department. John was appointed as a Materials Inspector in Glasgow for the Southern Railway, and then was appointed to the Locomotive Running Department, at Ryde, Isle of Wight, and as Locomotive Shed Master at Yeovil in Somerset and Brighton in Sussex. He retired from the post of Assistant District Motive Power Superintendent of the Central Division of the Southern Region of British Rail in 1964. His grandson, also Robert Urie, retired in 1999 from the privatised railway, his last role being Managing Director Trans Pennine Development of the company MTL.

Robert Urie survived many years in retirement, and he and his wife travelled across Canada by train. They moved home to Largs in Ayrshire, where Robert died on 6 January 1937, aged eighty-two.

Chapter 4

URIE'S H15S, DESIGN, CONSTRUCTION & EARLY OPERATION

Before his death in November 1912, Dugald Drummond had designed and obtained authority for a further series of 6ft wheeled 4-cylinder 4-6-0s, described in the August 1912 drawings as class K15. Drawings of a freight 0-8-0 class H15 had also been prepared, but the success of either was unlikely in that they perpetuated the saturated steam, long shallow grates and tortuous front ends that had been some of the factors leading to the comparative failure to meet expectations of his previous 4-6-0 designs.

Robert Urie's first decision on assuming responsibility at Eastleigh was to cancel the construction of these Drummond designs and obtain revised authority as early as April 1913 for ten 2-cylinder 4-6-0s to replace the authorised K15s.

These new engines differed radically from the Drummond predecessors, although some details were pure Drummond – cabs, dome covers and, initially, chimneys. The new engines were simple and robust, with Walschaerts external valve gear to the two 21in x 28in outside cylinders, 11in piston valves, and a sloping 9in grate with sufficient depth beneath the brick arch to allow for the provision of ample steam for prolonged periods. Bearings with ample surfaces were provided to overcome the previously all too common experience of Drummond's engines with hotboxes. Accessibility for maintenance purposes was greatly improved by a high-running plate over the cylinders and driving wheels, the latter having a small continuous splasher. An imposing round-topped boiler of 5ft 6in diameter and 13ft 9in length, with a grate area of 30sqft, and 180lb pressure, was constructed, fed by injectors and free of Drummond's complex cross-water tubes. The smokeboxes were of unusually large diameter, their appearance being accentuated by small smokebox doors.

Numbered 482–491, they cost £3,950 each compared with the estimated £3,700 of the proposed Drummond K15s and all had entered traffic before the outbreak of the First World War in August 1914. The cost included provision of superheating, although Urie decided to test this by providing 482-5 with the Schmidt equipment, 486-9 with the superheater designed by the Great Central Railway's J.G. Robinson and initially left 490-1 unsuperheated for comparative purposes. (The superheater royalties amounted only to £180-200 per locomotive.) The superheated engines weighed nearly 139 tons, including a large 57 ton 14cwt bogie tender, the saturated engines 136 ton 16cwt. The capacity of these large tenders was 5,200 gallons of water and 5 tons of coal.

The main frames were much stronger than on the earlier Drummond 4-6-0s and there was ample space for the axle and crank pin journals, reducing bearing problems, another source of trouble on earlier engines. The frames of the ten H15s lasted particularly well – when 30487 was stripped down for repair at Eastleigh Works in 1954, after forty years' operation, the frames showed few signs of their age. The result of the robustness of their design and ease of maintenance meant that the designer's objective of 100,000 miles between Works repairs was achieved. The new engines were turned out in full lined L&SWR passenger livery.

After the new engines had been run in on local passenger and freight services around Eastleigh, the majority were allocated to Nine Elms with just 488 and 489 going to Salisbury, but by March 1915 all ten were working from the London depot. They worked express passenger trains to Bournemouth

A colour postcard photo of the prototype January 1914-built H15, 486, in L&SWR livery, c January 1914. It was built with a Robinson superheater and did not receive its first General Overhaul until 1921, when it had amassed 228,971 miles in traffic. It was not withdrawn until July 1959, after a life of over forty-five years and mileage of 1.5 million. (MLS Collection)

and Salisbury by day and heavy fast freights to and from Southampton Docks or Salisbury by night and during this period of the war, also took charge of heavy troop trains.

A typical diagram worked by the H15s in 1914/5, was this Nine Elms turn:

10.15am Waterloo – Bournemouth Central express passenger train

3.44pm Bournemouth West – Waterloo express passenger

11pm Nine Elms Yard – Salisbury fast goods train

490 at Nine Elms in L&SWR livery 12 June 1920. 490 was one of two of the class that Urie built without superheaters for comparative purposes. This photograph was taken after its first General Overhaul in December 1919 when it was fitted with an Eastleigh superheater. (J.M. Bentley Collection/H.C. Casserley)

3am Salisbury – Nine Elms Yard fast goods

Another involved two return daytime express passenger turns between Waterloo and Salisbury and another, starting at Salisbury, included a passenger turn from Waterloo to Bournemouth and back before returning to Salisbury in the evening with the milk empties bound for the West Country. All expectations were being met, apart from a spate of hot boxes which Urie blamed on over-rigidity of the frame-staying, which was corrected at their first Works shopping. By September 1915, the ten engines had amassed on average 45,000 miles, three quarters on passenger work, and averaged coal consumption of around 45–50lb per mile, apart from the two saturated engines whose coal consumption was substantially higher (by 10lb per mile).

Despite this unfavourable comparison in fuel consumption, 490 and 491 displayed better acceleration in traffic, and both the Schmidt and Robinson superheaters used on the other engines developed different faults. Urie therefore designed his own variation, which became known as the Eastleigh superheater. Drummond's lone E14, No.335, had been laid aside for repair in 1912 and, because of its unsatisfactory performance, modifications had been envisaged. Urie left it alone until the success of his H15s had been evident. He then took the Drummond engine into Eastleigh Works in mid-1914 and completely rebuilt it to H15 specifications, with

Lone E14, No.335, in Southern Railway wartime plain black livery, built as a 4-cylinder locomotive by Drummond in 1907 and rebuilt as a 2-cylinder H15 by Urie between September and December 1914, at Nine Elms depot c1947.
(MLS Collection/J.D. Darby)

The rebuilt E14 under British Railways guise as H15 30335 at Eastleigh in 1957. It had received its last 'Intermediate' overhaul at Eastleigh in January 1955 and retained its large 4,500 gallon 'watercart' tender that it had acquired from the Drummond 4-2-2-0, No.720, in 1914.
(Colin Boocock)

two 21in x 28in outside cylinders and H15 pattern valve gear. The Drummond boiler was retained, but retubed and given a larger 31.5sqft grate area and 335 was the first locomotive to be provided with the new Eastleigh pattern superheater. Apart from the bogie wheels, boiler shell and tender, it was virtually a new engine. The total cost of the rebuild was £2,985, just under £1,000 less than building the new locomotives. Proving successful on 335, the Eastleigh superheater was subsequently fitted to all the H15s (and all Urie's later 4-6-0s and many Drummond classes). 490 and 491 were equipped with superheaters during their first Works overhaul visits in 1919.

Unlike the 482-491 series, 335 was initially painted the L&SWR dark green freight livery, as it was intended from the start for fast goods services. It was stationed at Eastleigh and worked Southampton Docks–Nine Elms freight trains and heavy troop specials. In mid-1915 it was reallocated to Nine Elms, but as it had retained the Drummond flat grate (although much larger) it was not as easy to fire as the new H15s and was consequently less popular. It was then reallocated to Salisbury where the local crews found it infinitely better than the F13s, G and P14s, with which they were then operating between Salisbury and Exeter.

During the war period the eleven H15s demonstrated their ability to handle every task offered with the minimum of attention. 482-491 averaged nearly 185,000

miles before receiving their first general Works overhaul and even 335 managed 104,000. These first heavy overhauls did not take place until 1917 and 486 managed to run until March 1921 before it needed serious attention. However, three of them were involved in collisions during the wartime period, 488 and 490 at Andover in 1914 and 1916, and 491 ran into the buffer-stops at Southampton. All of the incidents were blamed on the crews being unfamiliar with the locomotives' deceptively quiet running, causing speeds to be wrongly assessed. The most serious was when 488 ran into the back of a stationary goods train at nearly 50mph, but despite the huge damage to the freight vehicle, 488 demonstrated its robustness by incurring little damage and was

487 in Southern Railway livery, after overhaul in September 1925, built with Robinson superheater. 487 would be equipped with a Maunsell superheater and smoke deflectors at overhauls in 1929 and 1931 respectively.
(J.M. Bentley Collection)

483 at Nine Elms after the fitting of the SR standard smoke deflectors, October 1931. It was equipped with a stovepipe chimney in 1924 which it retained until 1943.
(J.M. Bentley Collection/Photomatic)

Urie H15 30485 at Eastleigh after a General Repair in March 1953. It was involved in a collision with a light engine, King Arthur 30783 *Sir Gillemere*, outside Bournemouth locomotive depot in 1955 and was then withdrawn from traffic. (John Scott-Morgan Collection)

able to be hauled to Eastleigh for repair.

At the Grouping in 1923, the H15s received the Southern Railway passenger green livery,

Urie H15 30482 at Eastleigh after withdrawal on 10 May 1959. 482 was built in March 1914 and its 5,200 gallon tender was exchanged with a standard Southern 5,000 gallon bogie tender at its General Overhaul in November 1954. At its withdrawal, it had 1.47 million miles to its credit. (Colin Boocock)

with the exception of 483, which for a short while was painted lined black, the freight livery. Urie's five-year plan that he'd presented to the L&SWR Board in 1919 had included additional H15 mixed traffic locomotives and ten of these had been authorised and were constructed in 1924 just after the Grouping, but with Urie N15 styled coned boilers and straight running plate clear of both cylinders and

driving wheels. These engines were fitted with stovepipe chimneys and Urie pattern 5,000 gallon tenders, similar to those of the N15s. Numbered 473-478 and 521-524, they entered traffic between February and September 1924 and cost, with the post-war inflation, over £10,000 each. Although they looked very different to the earlier Urie H15s, because of the boiler and smokebox variation, in fact they were still the Urie design, although constructed in the early years of the new CME, Richard Maunsell. Despite the boiler changes, there was little difference in performance, with similarity in both speed achievement and fuel consumption.

Meanwhile, the five Drummond F13 4-6-0s had languished, being seldom used even for freight traffic. Urie had intended to rebuild them as H15s, similar to his conversion of 335, but this had not been done by the Grouping. Maunsell agreed with Urie's intention, and took all five of the locomotives into Eastleigh Works in 1924 for rebuilding. The frames, cylinders, motion, coupled wheels and cabs were new, while

Official Works Grey photograph of newly built H15 No.476 at Eastleigh, March 1924, with stovepipe chimney. (MLS Collection)

No.521, the Maunsell-built H15 to the Urie design, with tapered boiler, after the replacement of the stovepipe chimney and the fitting of smoke deflectors, August 1931. 30521 was withdrawn in December 1961, one of the last four H15s all withdrawn that month (the others were 30475, 30476 and 30523). (J.M. Bentley Collection)

only the boiler shell, bogie wheels and tenders were retained from the F13s. They were considered in effect new engines, although their records going back to 1905 were kept as their history. They retained the flat Drummond firebox with the larger grate area as with 335, which made them less popular with firemen than the 482-491 series. They had large diameter smokeboxes similar to the new Urie engines, but had very squat stovepipe chimneys making their appearance seem even more massive. They had Maunsell Eastleigh superheaters, retained their original numbers, 330-334, and the cost of the rebuilding was £8,500, over £1,500 less than the newly built 1924 series. Their

Maunsell-built H15 to Urie design, 475, at Eastleigh after a visit to the Works when a new right hand cylinder was fitted, February 1947. It is still painted wartime plain black, dating from its overhaul in December 1943. (MLS Collection)

H15 30475 on a stopping train in the Bournemouth area, c1950. It had been renumbered, but this photo was taken before its Works visit in October 1952 and was painted in the BR mixed traffic lined black livery. (J.M. Bentley Collection)

H15 30476, after a light casual repair, stabled between a visiting Black 5 and a Urie H15 on Eastleigh shed, 7 September 1957. (John Hodge)

No.331 rebuilt by Maunsell from the Drummond F13 4-cylinder 4-6-0 in 1924, after overhaul at Eastleigh and the provision of straight sided smoke deflectors in November 1931. (J.M. Bentley Collection)

Drummond tenders that they retained had the water capacity raised from 4,000 to 4,300 gallons and the rebuilt engines weighed 129 tons 14cwt, the lesser weight being because of the smaller tender.

Maunsell had in fact authority to rebuild the G and P 14s as H15s also, but by the early 1920s the SR was believed to have sufficient 'mixed traffic' engines and the greater need was for capable express passenger locomotives and these Drummond 4-6-0s were converted by Maunsell as passenger

Rebuilt F13 No. 332, painted wartime plain black in 1942, and provided with a short flared chimney in place of its previous stovepipe in May 1947, seen here at Salisbury, its home depot, c1948. (MLS Collection)

Rebuilt F13 30334 at Eastleigh, waiting entry into Eastleigh Works for a General Overhaul at the end of 1952. It did not have its stovepipe chimney replaced by the flared variety until its next Works visit in January 1956. (MLS Collection)

Rebuilt F13 30331, thought to be passing Exmouth Junction with an Exeter – Salisbury stopping train, after its original 'watercart' tender had been replaced by the 5,000 gallon bogie tender from withdrawn Urie N15 30744 in March 1956, c1957. 30331 was withdrawn in March 1961, having run 1.15 million miles as an H15. (R.G. Jarvis)

N15s, as will be related in a later chapter.

330-334 were allocated to Salisbury to join 335, as this depot and its crews had greater experience of the flat-grated H15 and managed to cope with its steam-making ability through their skill and familiarity with the engine. Mostly they were used between Salisbury and Exeter on passenger and parcels services.

No spare boilers were built for the rebuilt Drummond engines, so they spent longer in Works at general repair time. The same was true initially for the Urie 482-491 series, but in 1927 Maunsell fitted 491 with a coned N15 boiler, thus releasing its Urie parallel boiler as a spare for the other locomotives of that series. 491 retained this type of boiler for the rest of its long life and had an excellent reputation. The Maunsell-built H15s to the Urie design were equipped with Urie N15 type boilers from the start and exchanged boilers at Works overhauls with both Urie and Maunsell N15 type boilers. A number of relatively minor changes

were made to the H15s in the early years, affecting their appearance. The most obvious was the replacement of the 'lipped' chimney with short stovepipe chimneys on some of the engines (483, 488 and 491). In the late 1920s the Maunsell-built H15s and 491 had their stovepipe chimneys replaced by Maunsell chimneys as applied to the Maunsell N15 King Arthurs. In following years all the H15s lost their stovepipe chimneys, carrying a flared but squatter type of chimney, similar to those of the new Lord Nelson 4-6-0s.

The most significant variation in appearance did not come until the

30331, one of the Salisbury rebuilt F13s, at Redhill, after working an engineering train to that location, 27 February 1960. (R.C. Riley)

30486, the first Urie H15 to be built at the end of 1913, at Nine Elms near the end of its forty-five-year-old life in 1959 – it was withdrawn in July that year having run 1.51 million miles. It had been based at Nine Elms throughout its career apart from the Second World War years during which it operated from Feltham. (MLS Collection)

The Urie 1914 H15 30491, subsequently provided with a taper boiler which it retained until its withdrawal in 1961, seen here on a local passenger train at Southampton Central, 17 January 1956. (Colin Boocock)

Maunsell H15 to Urie design, 30521, with 5,000 gallon bogie tender, at Nine Elms c1959. (John Scott-Morgan Collection/Photomatic)

Maunsell H15 to Urie design 30474 attached to a large 5,200 gallon capacity tender from a withdrawn Urie example, 30484, at Eastleigh, 9 December 1958.
(Colin Boocock)

late 1920s, when smoke deflectors began to be fitted to many of the Southern's larger locomotives. Forward visibility from these engines had been a problem for many years, with drifting smoke obscuring the driver's view and the Urie parallel boiler H15s all received the standard straight deflectors and the coned boiler engines, the standard SR inward-curved deflectors between 1929 and 1931.

By the early 1920s, immediately after the Grouping, the H15s were allocated to Nine Elms (482-491 plus 473-477of the 1924 engines), Eastleigh (478 and 521-524) and Salisbury (330-335). They were mainly employed on semi-fast passenger services (the Urie N15s had most of the express passenger services by then), van trains, and fast goods services between London, the West of England, Bournemouth and Portsmouth. In the summer months they were used on fast relief passenger services to Bournemouth and

Weymouth. They were capable of running heavy trains at 55-60mph on the level, but their valve gear limited their free running at higher speeds, although as 'mixed traffic' engines, they were useful on the summer relief services. Surprisingly the later Maunsell-built series were found initially to be inferior, suffering from

steaming problems, drifting smoke and steam. The replacement of the stovepipe chimneys helped the latter. Because of their weight and hammer blow on the track, the H15s were restricted to the main former L&SWR routes.

H15 30476 at Basingstoke awaiting right-away with a semi-fast passenger train from Basingstoke to Waterloo, c1959. It had been repainted during a Works visit in January 1958 and was withdrawn in December 1961, having run 1.26 million miles.
(R.C. Riley)

Chapter 5

URIE'S N15S, DESIGN, CONSTRUCTION AND EARLY OPERATION

The First World War years put many restrictions on the building and maintenance of locomotives at Eastleigh, and although Urie was aware of the L&SWR's desperate need for new passenger engines capable of hauling the heavier wartime loads and replacing the lugubrious and inefficient Drummond 4-6-0s, he was unable to construct any new design, although he had drawings available from 1916. However, by 1918, the Company's need was so apparent that he made a gang of men available to assemble ten new passenger 2-cylinder 4-6-0s that he had designed two years previously.

The first new engine, No. 736, appeared from Eastleigh Works in August 1918 at an assessed cost of £6,740. In many ways, 736 was similar to the designer's earlier H15, a simple and robust 2-cylinder engine with outside Walschaerts valve gear, although it had a tapered boiler, reducing the smokebox diameter and weight at the front end. This was the first use of such a boiler by the South Western, although unlike the Swindon practice, the taper was restricted to the front ring. 737 and 738 followed the same year and 739-745 in 1919. Subsequently a further ten were built in 1922 (746-752) and 1923 (753-755). Post-war inflation had increased the costs of these engines substantially, with 741 costing £7,765, 746 £8,237 and the final class members, £8,424, interestingly, appreciably more than the GWR's 1923-built Castles which, according to Swindon records, only cost £6,835 each. This is surprising and impossible to compare with any certainty as it is possible that different accounting methods make the figures skewed. One explanation may be the inclusion of the overheads of the Works costs. Investment in the Works at Eastleigh was much more recent than Swindon and costs of depreciation would be higher of the L&SWR facility. On the other hand the N15s were heavier with thicker frames, so the costs of the metal involved would have been higher and, of course, the N15 had a large bogie tender compared with the small standard Swindon 3,500 gallon 6-wheel tender with which the Castles were initially built.

The key dimensions of the new 4-6-0 were large outside cylinders 22in x 28in, coupled driving wheels of 6ft 7in diameter, boiler pressure of 180psi, a grate area of 30sqft and heating surface of 2,186sqft. The weight of the engine was 77 tons 17cwt and, with a bogie 5,000 gallon tender, 134 tons 18cwt. The running plate was straight above the cylinders with small continuous splashers over the coupled wheels. The engines were fitted with stovepipe chimneys. Tractive effort was 26,200lb.

736 was run in on local trains from Eastleigh to Bournemouth before being returned to the Paint Shop for application of

L&SWR's lined passenger livery, as introduced by Drummond, with elaborate lining and multi-striped boiler bands and brown borders on the tender as well as the black and white lining. On 10 September 1918, it went light engine from Eastleigh to Nine Elms for introduction into general service and immediately appeared at the head of the 11am Waterloo–West of England express, causing a sensation as its appearance was unheralded. 736-745 were initially all allocated to Nine Elms and worked the main express passenger trains to Salisbury and Bournemouth. After various tests with the new 4-6-0s and the best Drummond 4-4-0s (the D15s) in 1920, 740-742 were transferred to Exmouth Junction and 743-745 to Salisbury, leaving the D15s in charge of most of the Bournemouth services. The new locomotives soon acquitted themselves well on the switchback Salisbury–Exeter route, being allowed to haul 380 tons unaided. With an allowance of 470 tons east of Salisbury, the N15s fared less well, steaming problems on the long fairly level stretches of track becoming apparent all too frequently.

In 1919, main line speeds were still restricted to 60mph, but even so, some early logs recounted by Cecil J. Allen, O.S. Nock and others were not that impressive. 736 was timed over the first 50 miles from London on the 11 o'clock Waterloo at just over the hour, with no higher speed on the level than 56mph, needing 35% cut-off and full regulator to achieve this, even with only 350 tons behind the tender. West of Salisbury, where performance was

Brand new Urie N15 740 at Nine Elms in full passenger L&SWR livery, April 1919. It lasted thirty-six years, with only minor alteration (equipping with smoke deflectors the most visible) and was a post-Second World War oil-burner for a couple of years, being equipped from 1947 with electric generator operated headlamps. It was withdrawn in December 1955 and cut up at Brighton Works in June 1956. It ran 1,357,971 miles.
(John Scott-Morgan Collection)

Urie N15 741 *Joyous Gard* at its home depot, Nine Elms, c 1930. 741 was fitted with smoke deflectors in January 1928, and was withdrawn in February 1957, having run 1,386,007 miles. (John Scott-Morgan Collection/Photomatic)

notably better than the Drummond 4-6-0s, it was said that cut-offs of 50 or even 55% with full regulator were being used on the banks, not particularly economical working, despite the laudatory technical press of the day. However, Cecil J. Allen described a couple of runs between Salisbury and Exeter in 1920 and 1921 which contradicted this disappointing opinion. Both runs were made with the same engine and the same driver, 743 with Driver Charlie Clarke of Salisbury, the first when the overall 60mph line speed limit was in force. In the 1920 run, the 10.50am Waterloo had twelve coaches, 350 tons, with minimum speeds

of 34mph at Semley; 32½ before Templecombe; 21½ at Hewish; and 22 at Honiton but without the usual high speed impetus for the climbs. The following year with just ten coaches, 305 tons, he and 743 topped Semley at 40; 44½ at Sutton Bingham; and 33 at Hewish, with speeds in the mid-70s before most of the climbs. However, the train was brought to a stand at Seaton Junction by signals (a relief in front doing badly) and with full regulator and 55% cut-off accelerated and held 27½mph. Clarke had worked 743 relatively lightly before that and he and his mate, Fireman Searle, had achieved the economical working of just 30lb coal per mile

over a four-month period (including standing time and lighting up).

During this early period the N15s were each worked by a regular pair of crews and in consequence their annual mileage was averaging only about 35,000. However, in 1922 engine diagrams were tightened and through working to Exeter commenced with a crew change at Salisbury. Apparently this was not very successful and loss of time, mainly due to poor steaming, was common.

During the 1921 coal strike, 737 and 739 (the latter being the apparent 'black sheep' of the early engines) were converted to oil-burning. Both were then at Nine

Elms and worked to Salisbury and Bournemouth on normal express services, but combustion was unsatisfactory and led to many public complaints. Both were returned to orthodox coal firing by the end of the year. During the General Strike in 1926, both of these locomotives were again converted to oil-burners. Initially they worked parcels and freight duties around Eastleigh, though later they moved to Nine Elms for express passenger work. Both had been restored to coal burning by

Oil-burning 737 *King Uther* at Eastleigh during the General Strike of 1926. It was converted to burn oil between April and September 1921, and again from June to December in 1926. It ran approximately 19,000 miles in this form.
(John Scott-Morgan Collection/ Real Photographs)

Oil-burning N15 740 *Merlin* during the post-war conversion to this fuel due to the shortage of good steam coal, c1947.
(John Scott-Morgan Collection/ Real Photographs)

Oil-burning 740 *Merlin* at Eastleigh at the end of the 1946-1948 national programme, in the Bulleid sage green livery, c1948. (ColourRail)

Urie N15 30749 *Iseult* in the British Railways era at Basingstoke shed, 22 April 1956. The 1946 scheme oil-burning locomotives were equipped with electric headlamps, which they retained after conversion back to coal-burning. The generator can be seen behind the smoke deflector on the running plate. The loco is in the final Brunswick BR passenger green livery with the early small logo on the tender. (R.C .Riley)

December, six months later.

A final effort at oil-burning came after the Second World War when there was an acute coal shortage and the Government ordered a limited conversion programme. Ten N15s had been selected, but only 740/45/48/49/52 had been fitted with oil-burning equipment before the Government changed its policy, following pressure from the Treasury because of balance of payment problems and lack of foreign currency with which to purchase oil. The first conversion (740) was in December 1946 and the last to retain oil-burning equipment was 745, which was converted back to coal-burning by December 1948. These engines were also equipped with electric lights powered by Stone's turbo-generators, hidden behind the left-hand smoke deflector which replaced their oil headlamps and provided cab lighting, a system retained until their withdrawal.

By the Grouping, all the Urie N15s were completed, the last three still being painted in the L&SWR passenger livery. Most were still based at Nine Elms, with 743/44/45/50 at Salisbury and just 749 at Exmouth Junction. There were still complaints, however, about poor steaming and sluggishness of the N15s, especially east of Salisbury. Maunsell subjected 742 to a series of test runs in 1924-5 and on the first, in February 1924, steaming was so bad that pressure had fallen to just 120psi by Hook, between Woking and Basingstoke. The indicated horsepower did not exceed 950 at any point. 742 was returned to the Works and the steam passages were improved and valves altered to give better exhaust clearance. The second tests were in May 1924, but steaming was still poor and every time the engine was opened out, the boiler pressure dropped from 180 to 150-160psi. 742 was then fitted with a taller 'lipped' chimney and 5½in diameter blastpipe and on the third test run in April 1925 much better steaming was experienced. Steam pressure was generally maintained and an indicated horsepower rating of 1,250 was achieved without the engine being 'winded'. Coal consumption also improved as a result of the blastpipe modifications, estimated officially as a 17% reduction. Similar blastpipe modifications were made to the other nineteen N15s as they passed through Works for major

30749 *Iseult* at Eastleigh c 1955.
(MLS Collection)

repairs. Various further chimney/ blastpipe experiments continued, 755 being equipped for a while with a large diameter stovepipe with louvres. By 1928, all the Urie N15s had received the standard Maunsell flared chimneys as applied to his own N15 design.

In 1925, when the Maunsell N15s were under construction, the whole N15 class was given the names of King Arthur and his court, and the Urie N15s received some of the most attractive names with straight brass nameplates bolted to the continuous splashers over the centre driving wheels. The

brass lettering initially had a black background, though later this was painted red. The names allocated were:

736 *Excalibur*
737 *King Uther*
738 *King Pellinore*
739 *King Leodegrance*
740 *Merlin*
741 *Joyous Gard*
742 *Camelot*
743 *Lyonnesse*
744 *Maid of Astolat*
745 *Tintagel*
746 *Pendragon*
747 *Elaine*
748 *Vivien*

749 *Iseult*
750 *Morgan le Fay*
751 *Etarre*
752 *Linette*
753 *Melisande*
754 *The Green Knight*
755 *The Red Knight*

Despite the early favourable publicity surrounding the naming of the King Arthur class generally and the Urie N15s as well (and in the face of Maunsell's disapproval of the Urie engines being by inference equated with his own development), the publicity section's knowledge of the Arthurian legends creaked a little,

736 *Excalibur* in run-down condition in wartime black livery at the end of the Second World War, fitted with a Bulleid wide diameter multiple jet exhaust and straight smoke deflectors, c1945. 30736 was withdrawn in November 1956 from Bournemouth depot, having run over 1.45 million miles.
(R.C. Riley)

as two locomotives bore the name of the same young woman (*Maid of Astolat* and *Elaine*) who was of dubious virtue anyway, and 753's name, *Melisande*, cannot be found in any of the Arthurian legends at all, the confusion possibly caused by the opera/ballet Pelleas and Melisande as Sir Pelleas was an Arthurian Knight, the name being borne by Maunsell N15 778.

Problems also existed from the beginning with drifting smoke obscuring the drivers' vision – a similar problem experienced by the H15s – and various devices were tried in 1926-7, mostly on the Maunsell N15s, although 753 ran for a while with a curious small curved extension plate across the front of the smokebox. Maunsell derived his final solution after wind-tunnel tests at Eastleigh, and the Urie N15s were equipped with what became the standard SR design smoke deflectors at the end of the 1920s and early '30s.

Further modifications were made to the Urie N15 design by Maunsell. His N15 design had 20 ½in x 28in cylinders and the Urie engines' cylinders were lined up to 21in, reducing their tractive effort to 23,915lb as they went through Works after 1925, apart from 755 which retained its 22in diameter cylinders throughout its life. Six of these locomotives (740/43/45/46/48/52) were equipped with double exhaust ported steam valves between 1932 and 1934 and Bulleid fitted an improved version to 738/50/51/55 in 1941-2. Bulleid also fitted 736/37/41/52/55 with large diameter multiple jet exhausts. 755, with these modifications and retaining the

Urie N15 s747 *Elaine* at Nine Elms shed, immediately after nationalisation, before full renumbering, 1948.
(John Scott-Morgan Collection)

30740 *Merlin* at Bournemouth Central, in malachite green livery, renumbered but with 'Southern' still on tender, after return to coal-burning, but retaining electric lighting, c1949.
(MLS Collection)

in a link with the more powerful Lord Nelsons. Presumably the reduction in size of the Urie N15 cylinders was to reduce the chances of the cylinders beating the boiler and help cure their reputation for poor steaming, but with the other modifications made, 755 was able to maintain its performance with the larger cylinders. It was a pity that the Urie N15s were never given the long-travel valve events of the Maunsell engines, although money for investment in steam power was very limited on the Southern in the late 1920s, and the Bulleid modifications really came too late, and were again limited by other priorities in the middle of the Second World War. What could twenty 755 clones with long-travel valve gear have achieved? All speculation, I'm afraid.

A portrait of 30742 *Camelot*, May 1950, still painted in Malachite green, but renumbered and British Railways inscribed on the tender, as many N15s and King Arthurs ran in the immediate post-nationalisation period. (ColourRail)

22in x 28in cylinders, gained the reputation of being the outstanding member of the class (and the highest tractive effort at 26,245lb of any version of N15) and was used for several years

30739 *King Leodegrance* ex-works at Eastleigh in BR passenger green livery, with small early logo on tender, with an N15X stabled behind it, May 1952. (MLS Collection)

30750 *Morgan le Fay* ex-works after its last General Overhaul at Eastleigh, September 1955. 30750 was withdrawn just under two years later in July 1957, having run just under 1.3 million miles. (ColourRail)

30750 *Morgan le Fay* stored during the winter in the open at Nine Elms, January 1957. (MLS Collection)

30736 *Excalibur*, a Bournemouth-allocated Urie N15 for many years, fitted by Bulleid with wide diameter chimney and multiple jet exhaust, at its home depot in 1949. Still liveried in malachite green, it has been renumbered but still retains the 'Southern' legend on the tender. (ColourRail)

30752 *Linette* **with** wide diameter multiple jet exhaust at Bournemouth shed, renumbered but still in malachite green livery, with Drummond S11 4-4-0 30398 in the background, c1950. (L.M. Bentley Collection)

The 'super' N15, 755 *The Red Knight*, which retained its 22in cylinders and was fitted by Bulleid with large diameter multiple jet exhaust and improved double exhaust ported steam valves, at Nine Elms in April 1946. It was the first Urie N15 to be repainted in the Southern Railway malachite green livery after the war. It was operating in the Nine Elms Lord Nelson link at this time.
(MLS Collection)

755 *The Red Knight* at Nine Elms, c1947.
(R.C. Riley)

30755 *The Red Knight* at Basingstoke shed in 1956. It was withdrawn from this location in May 1957, having run 1.33 million miles. (R.C. Riley)

A colour photo of 30755 *The Red Knight* shortly before withdrawal at Basingstoke shed, 2 March 1957. (Ken Wightman)

Chapter 6

URIE'S S15S, DESIGN, CONSTRUCTION AND EARLY OPERATION

When Urie succeeded Drummond at Eastleigh he discovered that the latter's drawing office team had produced plans for two freight locomotive designs, a K15 4-6-0 and an H15 0-8-0. However, these designs had many features in common with his unsuccessful 4-6-0s and Urie hastily abandoned both ventures. Initially, he designed and constructed his own H15 4-6-0 as described in Chapter 4, but it was not long before he turned his thoughts to a freight engine version of this and his N15 passenger locomotive.

The L&SWR had been loaned ex-Great Central Robinson ROD 2-8-0s during the First World War for the additional freight traffic, especially the haulage of coal to Portsmouth and Southampton Docks from the South Wales coalfield. However, the amount of freight work suitable for their heavy haulage capability at slow speed was limited on the L&SWR routes, whose main freight traffic was general merchandise and perishables from the docks and West Country and required a freight locomotive capable of higher speeds. Urie had designed a 4-6-0 version of his N15 class with 5ft 7in coupled wheels back in 1916, but for the same reason – the lack of capacity at Eastleigh beyond locomotive heavy maintenance and war munitions construction – although authorised and ordered in 1916 and planned for completion by October 1918, no opportunity arose to build the prototype S15 before 1920.

Twenty of these locomotives, numbered 496-515, were constructed – 497-501 between February and June 1920, 502-506 between July and October, with 507-515, and lastly 496 (as the number 516 was already occupied by a Urie H16 4-6-2T), between November 1920 and May 1921. The average cost of each locomotive was £9,512, the first batch costing £8,405 and the last £10,230, demonstrating the rapid inflation that was taking place after the First World War. This seems expensive compared with the quoted costs of £5,826 for construction of the Swindon-built 1918/9 batch of 28XX 2-8-0 heavy freight locomotives, though less than the Government was asking for the redundant ROD 2-8-0s (£12,000). However, the RODs found no buyers at that price and ultimately the Government Ministry disposed of these 'second-hand' locomotives for just £2,500. Of course, differences in accounting practices cannot be ruled out and the GWR standard 3,500 gallon tender with which the 28XX were fitted was undoubtedly much cheaper than the Urie 5,000 gallon bogie tenders, the latter's additional capacity being necessary because of the absence of water troughs on the routes over which the S15s were required to work. The L&SWR 4-6-0s, being fitted with driving

508 at Strawberry Hill shortly after construction in L&SWR livery, c 1921. This engine suffered five serious hot box failures between February 1921 and March 1923, before its first General Overhaul in the autumn of 1925. (MLS Collection)

wheels a foot larger than those of the main heavy freight engines of the other railways, were not only designed for fast freight work,

but could also be used for troop train and relief passenger work during the increasingly heavy summer Saturday holiday traffic, so presumably the Company's Board were satisfied with the increased cost which provided greater versatility.

Many of their features were similar to those of the H15s and N15s – 2-cylinder outside Walschaerts valve gear 4-6-0s as a starting point. The cylinders were 21in diameter x 28in, coupled wheels 5ft 7in, boiler pressure 180psi, total heating surface

514 of the second batch of Urie S15s built in 1921, shortly after construction, in LS&WR goods green livery. The engine was not withdrawn until July 1963, after a life of over forty-two years during which time it accumulated 1.2 million miles, mostly on heavy freight work, based at Feltham for its whole life. (J.M. Bentley/O.L. Turner)

A different perspective of 508, posed when brand new, in December 1920. 30508 was withdrawn from Exmouth Junction in November 1963 – a paper transaction to the Western Region which never took place as it was withdrawn before a physical movement took place. Apart from three months spent at Exmouth Junction in 1929, this engine was based at Feltham throughout its life. (J.M. Bentley Collection)

2,186sqft and grate area 30sqft. Their weight in working order, with 5,000 gallon tenders, was 135 tons 1cwt. Tractive effort was 28,200lb. The boiler was similar to that of the N15s but was 4½in lower pitched than on the passenger engines. Stovepipe chimneys were fitted and the smokebox was shorter at exactly 5ft in length. Generous bearing surfaces were provided, with more than adequate lubrication and the design was robust as with Urie's other designs. The mechanical

parts were accessible for ease of maintenance, with a high running plate raised over the cylinders and clear of the motion.

The engines were superheated initially with Eastleigh superheaters, with Maunsell's own design replacing them between 1927 and 1932. The S15s suffered some of the same steaming problems as the N15s although the consequences on the traffic they worked was not so severe or obvious. Maunsell conducted blastpipe tests with 510

in 1931, which was given a King Arthur-style chimney as well as a modified blastpipe. The result was satisfactory, but with the lower pitched boiler, a taller chimney, as used on Maunsell's Moguls, was found more suitable. Boilers were interchangeable with the Urie N15s and the Maunsell-built H15s.

It is alleged, though not confirmed, that the first S15, 497, was put into traffic in full L&SWR passenger lined livery, but the remaining locomotives appeared in

A fine study of Urie S15 E501 at Nine Elms shortly after the Grouping in new Southern Railway lined black goods livery, February 1925. 501 was stationed at Nine Elms for exactly twenty years before joining its Feltham colleagues in 1941.
(John Scott-Morgan Collection/F. Moore)

498 with smoke deflectors, and in dark green livery, but minus the capuchon to the chimney, backing on to a freight at an unknown set of sidings, c1930.
(MLS Collection)

'goods green'. After the Grouping they were painted lined goods black, but this was later replaced by the passenger green livery because of their regular employment on passenger trains.

Initial allocation of these new freight locomotives was the first ten, 497-506, to Nine Elms and 507-515 to Strawberry Hill (Feltham), with the last one, 496, sent to Salisbury. Their immediate workload included heavy freights between Nine Elms and Eastleigh/Southampton Docks and some cross-London freights to Willesden and Reading. On summer Saturdays they appeared at the head of relief passenger trains to Salisbury and Bournemouth.

Tests were conducted between April and May 1924 and the S15s proved themselves more economical in service than the Maunsell 'N' and Billington 'K' Moguls. Maunsell was carrying out a number of tests at that time, determining his own design policy,

and the obvious superiority of the S15s over any other possible heavy freight rivals led Maunsell to use the Urie engine as the basis for his own freight design, incorporating some of the features he adopted for the 'improved' N15s. However, despite Urie's efforts to provide bearings with generous lubrication surfaces, the engines suffered from a number of early hotbox incidents,

507 paired with a Drummond 'watercart' tender, one of those provided to 504-510 to release the bogie tenders for the rebuilt N15Xs, at Fratton, 22 May 1937. It exchanged its 5,000 gallon bogie tender for the tender off Drummond 4-4-0 No.293 in August 1935 and retained it until October 1955 when it acquired a large 5,200 gallon bogie tender from withdrawn Urie H15 30483. (J.M. Bentley Collection/H.C. Casserley)

fifty-nine being recorded for the class of twenty locomotives before their first General Overhaul. Three of the class – 507, 508 and 511 – had

502, looking a little weary in wartime plain black livery, fitted with a Maunsell chimney in February 1947 similar to that used on the Maunsell U1 moguls, at Feltham, c1949. (MLS Collection)

An imposing shot of S15 30497 at Feltham in British Railways goods plain black livery c1955. 497 was one of the freight 4-6-0s loaned to the GWR for a couple of years, based at Old Oak Common for fourteen months from November 1941 and a few months at Tyseley in 1943. (MLS Collection)

as many as five hot boxes in the first couple of years and none of the class escaped without any incidents. This led to strenuous efforts to correct the problem and by the mid-1920s it had been eradicated as a major cause of concern.

As with the N15s, an example (the new 515) was converted to oil-burning at the onset of the 1921 coal strike, using the same 'Scarab' patent oil-burning equipment. The freight engine was used on both passenger and parcels trains,

including the 11.30am Waterloo–Bournemouth (an Eastleigh diagram) and a comparison made with the coal-burning 510. Although performance was satisfactory, the cost of the fuel used per mile was very nearly twice that of 510, and 515 was converted to coal-burning by October 1921. Again, as with the two N15s, 515 was again an oil-burner during the 1926 General Strike and was used to work imported coal from Poland from Southampton Docks to Reading.

Like other L&SWR and Southern locomotives, drifting smoke was a problem, not solved until the SR's standard smoke deflectors were fitted in 1929-30. Although all were fitted initially with bogie 5,000 gallon tenders, in 1935, when such tenders were required by the newly rebuilt N15X class (see chapter 8.3), seven S15s (504-510) exchanged theirs for the Drummond 4,000 gallon variety off withdrawn 4-4-0s. Later, many received large tenders again, as

30512 at Eastleigh waiting Works attention alongside H15 30330 and S15 30497 behind, c1954. 30512 was the last survivor of the class, being withdrawn in April 1964 and sold to Woodham Brothers scrapyard at Barry. It was however cut up in January 1965 – one of the few engines sold to Woodhams that did not get purchased by the railway heritage industry. It was based at Feltham throughout its forty-four year life. (MLS Collection)

30510 ex-works at Eastleigh with an M7 awaiting a decision on its fate behind and Bulleid Pacific 34090 *Sir Eustace Missenden*, before rebuilding, in the background, April 1959. (R.C.Riley)

during the 1930s much 'swapping' of tender types between all classes was occurring. When the Urie S15s were built, no spare boilers were made and Maunsell ordered eight N15 boilers from the North British Locomotive Company in December 1924 to be used for Urie H15s, N15s and S15s and also for his own designs. The first Urie S15 to receive one of these boilers was 497 in February 1928 – the boiler was pressed at 200psi and had Ross pop safety valves, smokebox door-shaped as for other Maunsell engines, Maunsell superheater and snifting valves.

The last Urie S15 to be built, although the first numerically, 30496 in British Railways days, but unaltered from its initial construction apart from chimney and smoke deflectors, with a BR Standard 82XXX 3MT tank engine at Eastleigh, c1960. (Ken Wightman)

Urie Goods S15 4-6-0 30500 in BR freight black livery, at Eastleigh, 18 May 1963. (R.C. Riley)

The preserved Urie S15 30506 seen here at Brighton in BR days in the goods black livery, c1962. It is now being overhauled on the Mid Hants 'Watercress Line' after fourteen years' operation in SR green livery between 1987 and 2001, and was anticipated to return to traffic in 2015. (Ken Wightman)

30512, the last surviving Urie S15 in British Railways operation, at Eastleigh after withdrawal the previous month, 18 April 1964. Behind it stands 'Battle of Britain' 34075 264 *Squadron* withdrawn the same month. (David Clark)

Chapter 7

RICHARD EDWARD LLOYD MAUNSELL

Richard Maunsell was born at Raheny, County Dublin, in Ireland on 26 May 1868. His predecessors were land-owners and were in the legal profession, but from an early age the boy showed his primary interest to be engineering. He was one of a large family – six brothers and four sisters – and he attended a Public School, the Royal School at Armagh, in 1882, before training, after pressure from his father, for a Law degree at Trinity College, Dublin, in 1886. However, the backbone BA course at the university was followed by all students and he was able to specialise in engineering, and was also to benefit from his solicitor father's contacts with the Board of the Irish Great Southern & Western Railway, becoming simultaneously a pupil of H.A. Ivatt at its Inchicore Works.

Maunsell managed to combine the continuation of his Trinity college studies with his Inchicore apprenticeship, as well as becoming an outstanding sportsman in both athletics and cricket, which he played for his college. In 1891 he completed his BA course at Trinity and left Inchicore for a year's experience under John Aspinall of the Lancashire & Yorkshire Railway, the result of a close Ivatt-Aspinall relationship. Clearly the engineering establishment already had the view that Richard Maunsell was a potential high-flyer. Maunsell worked at the newly opened Horwich Works where Aspinall's successful 2-4-2Ts were currently under construction. Alongside Maunsell were three other young men who later became CMEs, Henry Fowler, Henry Hoy and George Hughes, and Maunsell retained contact with these later, especially Henry Fowler after they'd both become CMEs of two of the 'Big Four' in 1923.

Maunsell had a number of basic shed appointments in the Blackpool and Fleetwood District after experience on the design side, and during this time, was courting Miss Edith Pearson whom he'd met during his contacts with the Aspinall family. However, Edith's father prevented their engagement until Maunsell was better able to assure him of his career earning prospects, so Maunsell sought a higher paid post and successfully applied to be Assistant District Locomotive Superintendent of the East India Railway based at Jamalpur at a salary of Rs480 a month (approximately £323 a year) which was nearly double his salary on the L&YR. The East India Railway was very extensive, the second largest railway in India. Maunsell gained rapid promotion there, and after a spell at Tundla on the Allahabad–Delhi route, was transferred back to Jamalpur, then as the Principal District Locomotive Superintendent.

However, even the increased salaries available to Maunsell were insufficient to persuade Edith's father to endorse the proposed marriage. H.A. Ivatt was appointed to the senior engineering post of the Great Northern Railway when Patrick Stirling died. His former Assistant, Robert Coey, then took up the vacant Superintendent's post of Ireland's GS&WR, which paved the way for Maunsell to be appointed at the young age of twenty-eight to the post of Assistant Locomotive Engineer and Works

Richard Maunsell at Ashford, c1914.
(G.M. Rial)

Manager in March 1896, and Richard and Edith Pearson were married in June of that year.

Maunsell immediately set about the reorganisation and modernisation of Inchicore Works and between 1897 and 1902 the GS&WR increased its network size by 70% through company take-overs, with Inchicore becoming responsible for the replacement and maintenance of the increased locomotive fleet. Maunsell was a close observer and protagonist of Coey's locomotive developments in the first decade of the new century, including 4-4-0s and 2-6-0s, and experiments with superheating, until Coey was forced to resign in 1911 because of increasing ill health. Richard Maunsell was appointed Locomotive Superintendent on 30 June 1911. Edward Watson was appointed to Maunsell's previous post, after having been Assistant Works Manager at Swindon, from where he brought knowledge of the developments being pioneered by Churchward. Maunsell designed just one tender 4-4-0 at Inchicore in 1912, a development of a Coey-initiated design, but with a Belpaire firebox and a higher boiler pressure.

Maunsell's reign at Inchicore was short-lived, for in 1913 he was approached by the South Eastern & Chatham railway seeking a replacement for their retiring Locomotive, Carriage and Wagon Engineer, Harry Wainwright. He was appointed in December 1913 and had the immediate task of reorganising Ashford Works which could not cope with the workload then placed on it. Maunsell brought his key skills of administration and efficient management, and his engineering knowledge required to tackle the need for greater power to manage the increasing passenger loadings on the SE&CR. The immediate need was met by the already ordered 'L' class 4-4-0s, ten of which were supplied by Borsig's of Berlin and twelve direct from Beyer Peacock's Works in Manchester, to alleviate the shortcomings at Ashford.

Maunsell soon assembled a very competent team, including James Clayton from Derby and George Pearson and Harry Holcroft from Swindon, although the onset of the First World War restricted their immediate influence. Some rebuilding of older classes took place, mainly reboilering, but in 1914 the Government created the Railway Executive Committee to take charge of the railways during the wartime period, and Richard Maunsell was appointed as Chief Mechanical Engineer to this body. Some of his work involved the overseeing of maintenance of locomotives in Belgium and Northern France, working under ROD auspices, and at the end of the war he was awarded the CBE for his services. During this time, however, he still found the time to design his prototype locomotives for the SE&CR, the 'N' class Mogul and the 'River' class 2-6-4 express passenger tank engine.

In the period between the end of the war and the Grouping, Maunsell set about rebuilding a number of the Wainwright 'D' and 'E' 4-4-0s with 10in piston valves, long travel gear and superheaters. Twenty-one Ds and eleven Es became D1 and E1 classes and were so successful that they lasted until the early 1960s. Maunsell had been poised to act as the CME of the proposed nationalised railway after the war, but political views changed and the Grouping proposed by Sir Eric Geddes, then Minister of Transport, came about under the Railways Act of 1921, implemented on 1 January 1923.

With Robert Urie's retirement at age sixty-eight, Maunsell was the natural successor as CME of the new Southern Railway, inheriting a fleet of 2,285 steam engines of 115 different classes, with little standardisation. Richard Maunsell had absorbed some of the best developments from Churchward at Swindon – taper boilers, top feed, Belpaire fireboxes, long-travel, long-lap valve gear – which had already proved their value in the Moguls and rebuilt 4-4-0s, then turned his attention to the larger locomotives he'd inherited from Urie on the L&SWR. He went on to develop the N15 and S15 classes (see later), as well as his own 4-cylinder Lord Nelsons and 3-cylinder masterpieces, the V Schools class.

Richard Maunsell was a consummate and skilled manager and administrator, and popular with his team and staff. His influence in new steam engine design was circumscribed by the Southern Board's priority of investment in electrification, restricting money available for the wholesale standardisation of the steam stock as happened on the GWR under Churchward and Collett, and Stanier on the LMS, so older designs were retained in addition to the basic N15, H15 and S15 family which were the nearest the SR got to standardisation in this inter-war period.

Increasing ill health caused Richard Maunsell to take retirement in 1937, when he was sixty-nine years old. He formally handed over responsibility for the SR Motive Power Department to Oliver Bulleid on 31 October, having had a strong and successful relationship for many years with the Southern's General Manager, Sir Herbert Walker. He left the railway with 1,852 steam engines of 77 classes and a substantially electrified network, retiring to spend his days involved in the life of his local Parish Church, which he and Edith attended regularly in Ashford.

He died in March 1944, leaving his wife and married daughter, his only child, and is buried in Bybrook Cemetery, Ashford, a few hundred yards from his house, 'Northbrooke', where he had lived for thirty-two years.

Chapter 8

RICHARD MAUNSELL'S DESIGNS

8.1 N15s: design, construction and early operation

On taking over the management of the new Southern Railway's locomotive and carriage policies and practice, Richard Maunsell was faced with the Company's shortage of capable express passenger power for both the former L&SWR main lines and the increasing importance of the continental boat train traffic to Dover and Folkestone Harbours. Drummond's 4-6-0s, with the exception of the moderately successful T14s, had already been rebuilt or relegated to occasional freight work. The twenty Urie N15s, despite the publicity, were not living up to their perceived reputation and were not in the same league as the GWR's 4-6-0 fleet and the LNER's new Pacifics. The Bournemouth and Portsmouth main expresses, and the Eastern Section's boat trains, were in the hands of capable 4-4-0s, but post-war traffic, and therefore loads, were increasing.

Maunsell and his team were well aware of developments in boiler design, modern front end and long-travel valve gear then being exploited to good effect at Swindon

The prototype theoretical rebuild of Drummond G14, 453 *King Arthur*, at Waterloo station, February 1925. 453 had a thirty-six-and-a-half-year existence in which it clocked up 1,606,482 miles, the highest mileage of any King Arthur. The Salisbury based King Arthurs established higher mileages on the whole, compared with the 'Scotch Arthurs' because of their regular use on the West of England route.
(J.M. Bentley Collection)

Drummond P14 449, renumbered 0449 to make way for Maunsell's new Eastleigh Arthur. This 4-6-0, allegedly rebuilt as 449 *Sir Torre*, was retained to act as a test bed for some of Maunsell's design ideas for the Lord Nelson class. It was scrapped at the end of the tests, and the tender attached to the new engine. (J.M. Bentley Collection/H.C. Casserley)

and he already had successful experience in the design and operation of his 'N' class Moguls and the rebuilding of the Wainwright 'D' and 'E' 4-4-0s. He therefore took the Urie 4-6-0 as a starting point and applied his knowledge and experience to produce an 'improved'

N15, with a higher 200psi boiler pressure, valve travel in full gear increased from 5⅛in to 6 9/16in with 1½ in lap instead of 1in, an increase in the superheater surface and improved steam passages and blastpipe arrangements. The ash pan was redesigned to provide

more air to the grate and a flared chimney replaced the stovepipes that the Urie engines still bore at that stage. The balancing of the weights was also improved to reduce the heavy hammer blow on the track that the Urie H15s and N15s delivered. The new engines were to be fitted with Maunsell's own design of superheater. Lessons from Maunsell's tests of the Urie N15, 742, recounted in Chapter 5, were absorbed, and the draughting, blastpipe and chimney design reflected these.

An immediate authority to rebuild the ten G14 and P14 Drummond 4-6-0s was granted, at a cost of £6,320 each, although in reality very little of the Drummond engines was retained in the reconstruction. One of the P14s, 449, gave the game away, as it was retained for experimentation of the eight exhaust beats per wheel revolution being considered for the Lord Nelsons then at the early design stage. Renumbered 0449, it ran for several months at the same time as its alleged rebuild, N15 449, was running in. For the other nine engines, the main part retained was the Drummond eight-wheel

Eastleigh Arthur 449 *Sir Torre* as built, c1925. It was withdrawn as 30449, with bogie tender, from Salisbury shed in December 1959, having run over 1.37 million miles. (Colin Garratt Collection/Rev A.W. Mace)

4,300 gallon and 5ton coal capacity 'water-cart' tender with inside frames and possibly minor parts such as the bogie wheels.

Other key dimensions of the new engines were grate area 30sqft (same as the Urie N15s), total heating surface of 2,215sqft, slightly reduced diameter cylinders of 20½in x 28in stroke, improved shape and size of the main steam pipes to the cylinders, with a total engine weight of 80 tons 19cwt. Tractive effort was slightly reduced, despite the higher boiler pressure, on account of the reduced cylinder size, being 25,320lb at 85% working pressure. The cab design of these ten so-called rebuilds was similar to the Urie N15s, an outline that restricted their use to the former L&SWR main lines. The 453-457 G14s were rebuilt first, 453 appearing in traffic in February

1925. These five engines were completed by March and the P14 448-452 series by June 1925. 449, newly painted in SR livery and coupled to a Urie bogie tender, was exhibited in May that year at the Stockton & Darlington Centenary celebrations.

The names of the ten Eastleigh Arthurs were:

448	*Sir Tristram*
449	*Sir Torre*
450	*Sir Kay*
451	*Sir Lamorak*
452	*Sir Meliagrance*
453	*King Arthur*
454	*Queen Guinevere*
455	*Sir Lancelot*
456	*Sir Galahad*
457	*Sir Bedivere*

Maunsell carried out tests with the new 451, similar to those conducted with 742, and an immediate

449 *Sir Torre*, being inspected by depot staff, c1925.
(Colin Garratt Collection/Rev A.W. Mace)

The Eastleigh Arthur, 449 *Sir Torre*, allegedly rebuilt from P14 0449, at Eastleigh after fitting with smoke deflectors, April 1929. Behind is a Drummond 4-4-0 and an 'N' class mogul.
(J.M. Bentley Collection)

improvement in the engine's effectiveness was noted, with power output of nearly 1,500 horsepower being achieved. The *Atlantic Coast Express* was used for the tests with a 14-coach load weighing 440 tons tare. The schedule was 93 minutes and the engine was worked on full regulator almost throughout with cut-off varying between 20-25%. Steam pressure was well maintained and coal consumption was no higher than 42lb per mile, giving coal consumption per indicated horsepower hour almost as good as the best results being recorded by the GWR's new Castles. Salisbury received the

453 *King Arthur* fitted with experimental small side smoke deflectors, c1927. (J.M. Bentley Collection)

Rebuilt P14 448, *Sir Tristram*, after fitting with the standard SR smoke deflectors, at Salisbury, c 1929. (J.M. Bentley Collection)

'**Rebuilt' P14,** Eastleigh Arthur No.450 *Sir Kay* approaches Clapham Junction with the 11am Waterloo-West of England Atlantic Coast Express, c1925. (MLS Collection)

first rebuilds, 453-457, to work in particular heavy expresses over the switchback Salisbury–Exeter route, 448 and 449 went to Exmouth Junction and the last three, 450-452, were allocated to Nine Elms. The initial experience with these engines met expectations.

Eastleigh's capacity was limited at that time to the building of the ten H15s with tapered boilers, the completion of the last three Urie N15s and the building of the new N15s from the carcases of the G and P14s. With its normal maintenance work and the development of the new Lord Nelsons as well, the urgent need for more express engines was met by a contract with the North British Locomotive Company in Glasgow to construct twenty new 4-6-0s to the Maunsell N15 design, at an agreed price

Another alleged 'rebuilt' P14, 452 *Sir Meliagrance*, at speed between Walton-on-Thames and Weybridge with the 11am Waterloo *Atlantic Coast Express* with new Maunsell carriage stock, c1926. (MLS Collection/Real Photographs)

771 *Sir Sagramore* shortly after delivery in 1925 on the turntable at Dover. Dover Castle can be seen in the background. (MLS Collection)

of £7,780 each, or £10,085 with the 5,000 gallon bogie tender, as fitted to the Urie N15s. There were a few important detailed differences, the main one being the shape of the curved cab to enable the engines to work on the South Eastern Section. These locomotives were delivered from May 1925 onwards, and a further ten were added, at the same contract price. The ten Eastleigh-built engines were known, fairly obviously, as 'Eastleigh Arthurs' with the Glasgow engines being nicknamed (somewhat ungrammatically) as the 'Scotchmen', or 'Scotch Arthurs', as by this time John Elliot, the new PR manager, had devised the happy idea of naming the whole class after King Arthur and the Knights of the Round Table.

The first twenty Scotch Arthurs received between May and July 1925 were:

763	*Sir Bors de Ganis*
764	*Sir Gawain*
765	*Sir Gareth*
766	*Sir Geraint*
767	*Sir Valence*
768	*Sir Balin*
769	*Sir Balan*
770	*Sir Prianius*
771	*Sir Sagramore*
772	*Sir Percivale*
773	*Sir Lavaine*
774	*Sir Gaheris*
775	*Sir Agravaine*
776	*Sir Galagars*
777	*Sir Lamiel*
778	*Sir Pelleas*
779	*Sir Colgrevance*
780	*Sir Persant*
781	*Sir Aglovale*
782	*Sir Brian*

Despite initial disappointment in the Scotch engines' performance, especially on the continental boat trains, the order was increased to thirty engines and the following were delivered to the Southern between August and October 1925:

783 *Sir Gillemere*
784 *Sir Nerovens*
785 *Sir Mador de la Porte*
786 *Sir Lionel*
787 *Sir Menadeuke*
788 *Sir Urre of the Mount*
789 *Sir Guy*
790 *Sir Villiers*
791 *Sir Uwaine*
792 *Sir Hervis de Revel*

As well as poor performance on the road, these engines were initially unreliable and the Southern General Manager called for a special report in February 1926, which highlighted a number of faults in the construction process at the North British Works, possibly caused by corners being cut to save costs – the profit margin was said to be less than £10 per locomotive. Also, performance on the Eastern Section, because of the drivers' method of operating the engines, may have been at fault. Western Section drivers tended to use short cut-offs with a fully opened regulator, whereas Eastern Section drivers more frequently used long cut-offs with partially opened regulators. Maunsell therefore initiated a further set of tests with 768 on both routes with a SE driver and 778 with a SW man, their regular drivers using the methods they normally employed. Whilst timekeeping on all the tests was excellent, the fuel economy of 778 and the South Western driver was substantially better by

New Scotchman 764 *Sir Gawain* on a continental boat train bound for Dover, c1925. 764 was the only Scotch Arthur to achieve less than a million miles in its life, having spent most of its career on the Eastern Section with a shorter post-war allocation to Bournemouth, mainly for the cross-country trains.
(MLS Collection)

5-10lb of coal per mile, and Eastern Section locomotive inspectors were instructed to counsel their drivers accordingly.

There may have been further reasons for the initial unexciting performance on the road. The state of the track on both the former

The first Scotchman, 763 *Sir Bors de Ganis*, on a down continental express, c1927.
(Colin Garratt Collection/Rev A.W. Mace)

800 *Sir Meleaus de Lile* after fitting with the standard smoke deflectors, November 1927. 800 was one of this batch that failed to reach the million mile mark, achieving just 960,500 miles before its withdrawal from Eastleigh in September 1961. (MLS Collection)

SE&CR and L&SWR routes was not up to the standard of elsewhere and, after the Sevenoaks derailment of the River class tank engine at speed, the Chief Inspecting Officer of Railways set up a series of tests with a number of different classes of locomotives on both routes. One of the assessors supporting the inspector was Nigel Gresley, who was very critical of the riding of the King Arthur on the South Western Section. Early logs by Cecil J. Allen and O.S. Nock reported disappointing downhill speeds (little over 70mph) between Basingstoke and Winchester and on the level between Basingstoke and Waterloo and this could well have been because of the reluctance of drivers to run hard over inferior track. Apparently the N15 on

test was reported as rolling quite severely at speed.

Cecil J. Allen had footplate runs in October 1925 through from Waterloo to Exeter on the Atlantic Coast Express behind 774 *Sir Gaheris* to Salisbury with ten coaches, 305 tons, reduced to eight from Salisbury, 245 tons, with 451 *Sir Lamorak*. Driver Barber of Nine Elms worked with full regulator and 25% cut-off, 20% from New Malden producing speeds in the mid-60s to Basingstoke, with short bursts of 25%, and 80 down Porton Bank, reaching Salisbury in 89½ minutes (86¾ net). Exmouth Junction's Driver Burridge on 451 for the most part used 20% cut-off on the level and downhill and 30% on the uphill stretches, varying the regulator opening. Just 30% on

Honiton produced virtually 30mph at the summit and with no speeds higher than 77½mph (reached several times), got to Exeter in a few seconds under 96 minutes (94½ minutes net). Nothing spectacular, but efficient and economical enough working. He did not comment on the riding and assumption must be made that the running was not inhibited by either rough-riding or drifting smoke.

The combination of the changed driving techniques and the correction of the construction flaws at Eastleigh (763/64/67/69/70 /73/75/78/90/92 were the main offenders and all visited the SR Works before the end of the year) gave the SR management sufficient confidence to order further locomotives for the Brighton ex-LB&SCR route, but equipped with six-wheel 3,500 gallon tenders to enable the engines to fit on the shorter wheelbase turntables on that route, most of whose previous express locomotives had been tank engines. In fact a further twenty-five, to be 793-817, had been authorised, but only fourteen were built, the last sixteen being replaced by Lord Nelsons 850-865, the design of which by mid-1926 had been completed. 4,000 gallon six-wheeled tenders intended for some of this batch were then allocated to the Eastern Section allocated engines, 763-772. The names and numbers of the 1926 N15s built at Eastleigh were:

793	*Sir Ontzlake*
794	*Sir Ector de Maris*
795	*Sir Dinadan*
796	*Sir Dodinas le Savage*
797	*Sir Blamor de Ganis*
798	*Sir Hectimere*
799	*Sir Ironside*
800	*Sir Meleaus de Lile*
801	*Sir Meliot de Logres*
802	*Sir Durnore*
803	*Sir Harry le Fise Lake*
804	*Sir Cador of Cornwall*
805	*Sir Constantine*
806	*Sir Galleron*

The improved availability and performance of the Scotch Arthurs and the allocation of 773 -782 to Nine Elms and the last ten 1925-built engines to Bournemouth, caused a reallocation of the Urie N15s, 737/39/40/47 to Exmouth Junction and 754 to Salisbury, so that all the main expresses west of Salisbury were now in the hands of Urie or Maunsell N15s. The Central Section engines, 793-795, were prevented for a few weeks for entering their intended service

as they had not been cleared for operation on the Brighton lines out of Victoria, but they were allocated to Brighton in July 1926 and joined at Brighton and Battersea by the newly built 796-806. They immediately displaced the River tanks and the H2 Marsh Atlantics on the Victoria–Brighton and some off-peak Eastbourne services, where they gave excellent services until electrification in 1933, although it took a little time for the former LB&SC drivers to get used to them and the best methods of working, after being so used to their Pacific and Baltic tanks.

The N15s were involved in a complex series of tender changes in their first few years. Initially, the Eastleigh Arthurs retained the Drummond 'watercart' tenders from the G and P14s and, in fact,

kept them until the 1950s when they received bogie 5,000 gallon tenders from withdrawn Urie N15s with 457 receiving the large bogie tender from withdrawn Urie H15 490. However, the Scotch Arthurs were different. 763-792 were all built with Urie double-bogie 5,000 gallon tenders, weighing 57tons 11cwt. Because of the shorter distances involved, the Scotchmen allocated to the South Eastern Section (763-772) were allocated new 'Ashford pattern' 4,000 gallon six-wheeled tenders between 1928 and 1930, apart from 769, which was reassigned to the South Western Section at this time. Then, in 1930, 768 and 770-772 received 5,000 gallon bogie tenders from new Maunsell S15s when it was found necessary to loan Stewarts Lane engines to Nine Elms, and

764 *Sir Gawain* with six-wheel tender leaves the Shakespeare Cliff Tunnel with a Boat Train for Dover Marine, c 1930. 764 exchanged the 5,000 gallon bogie tender with which it was built for a six-wheel tender at its first Works overhaul in April 1928, and regained a bogie tender in 1936.
(J.M. Bentley Collection/F. Moore)

the Brighton electrification) became much more flexible as the 793-806 series, built with six-wheel 3,500 gallon tenders, were sufficient for Eastern Section duties. This series retained these tenders until the Kent Coast electrification in 1959, when some were transferred to the South Western Section and 30793/95/96/98/800/02/03/06 received bogie tenders off withdrawn N15s.

The allocation of the Maunsell N15s in numerical blocks to the same shed may have had administrative advantages, but it meant that engines at the same shed ran up their mileage at a similar rate and would be requiring Works maintenance at around the same time. This led to engines being borrowed from other sheds, especially during the summer holiday period and the need to temporarily exchange tenders to ensure engines from the Eastern Section with six-wheel tenders were capable of use on the longer South Western routes. If the

Six-wheel tender Arthur 796 *Sir Dodinas le Savage* departs London Bridge with a train for Brighton, c1926. 796 retained this tender with which it was built until its very last Works visit in January 1961, when it received the bogie tender from the withdrawn 30766. (J.M. Bentley Collection/Real Photographs)

later, in 1936-7, the five other Eastern Section engines, 763-767, received bogie tenders off S15s 833-837 (as the S15s required shorter wheelbase tenders for operating over the Brighton Section). At one

stage, 768 acquired a large eight-wheeled flush-sided tender off a Lord Nelson and worked for several years on the Salisbury–Exeter route. The allocation of the Scotchmen between all three Sections (after

'Scotch Arthur' 766 *Sir Geraint*, within a month of its delivery from the North British Works, passes Bromley with a Pullman Boat Train bound for Dover Marine, before the naming of the train as the *Golden Arrow*, May 1925. (J.M.Bentley Collection)

771 *Sir Sagramore*, delivered from North British in June 1925, departs from the quayside at Dover Marine with a boat train for Victoria, c1925.
(J.M. Bentley Collection)

765 *Sir Gareth* passes Sydenham Hill with the 'Down Continental' boat train, September 1927.
(J.M. Bentley Collection/Real Photographs)

Eastern Section required additional locomotives, Nine Elms invariably loaned members of the Lord Nelson class rather than the N15s which they retained for the Salisbury and Exeter route in particular.

In the late 1920s until the electrification of the Brighton line, the N15 allocation settled down as: Nine Elms, 450-452, 773-782; Salisbury 453-457; Exmouth Junction 448, 449; Bournemouth 783-792; Battersea (Stewarts Lane) 763-772 ; Brighton 793-806. The latter were used on the Brighton and Eastbourne services, the Stewarts Lane engines worked to Dover, Folkestone and Margate.

A further problem to be addressed before the 1920s were over was that of drifting smoke obscuring the driver's view, another possible reason for the

767 *Sir Valence* climbing Grosvenor Bank with the 10.35am Continental Pullman train that became the *Golden Arrow* in 1929, c1927. The Nord Railway of France had designated the Calais–Paris Pullman train as the *Fleche d'or* in 1926.
(John Scott-Morgan Collection)

ineffective, 772 was equipped in September with large straight-sided metal plates as utilised by the Standard German Railways '01' Pacifics, which solved the problem but did not enhance the appearance of the engine. 783 was fitted with a shovel-shaped device round the chimney which looked even worse. Despite the appearance, the DR '01' deflectors were the most effective and efforts were made to use similar smaller and more aesthetically pleasing designs. In 1927, 453 received small side smoke deflectors which reached just up to the mid-point of the smokebox. The ultimate successful plates took the inwardly curved shape of the 453 example with the height of 772's German design and were fitted to most Maunsell and Urie N15s by 1928. However,

early restrained running. This problem was common to all the Urie and Maunsell 4-6-0s and a first experiment to try to solve the problem occurred as early as February 1926 when small metal wings were fixed behind the chimney of 450. As this was

Six-wheel tender Arthur, 796 *Sir Dodinas le Savage*, at Stewarts Lane, after repainting in malachite green in 1946. New Battle of Britain, 21C163 229 *Squadron* is in the background, c1947.
(MLS Collection)

Six-wheel tender Arthur 794 *Sir Ector de Maris* with a standard Maunsell chimney without the capuchin, at Stewarts Lane, c1935. 794, having spent the early part of its life restricted very much to the Brighton line, was withdrawn from Basingstoke as late as August 1960 with only 903,663 miles to its credit, the lowest mileage achieved by any King Arthur. (John Scott-Morgan Collection/ Photomatic)

there were still efforts to avoid the use of side smoke deflectors and 773 was fitted in 1930 with a semi-circular front projection at the top of the smokebox, but this was not continued. On the LMS, the similar high-pitched boiler Royal Scots with shorter chimneys were having even more smoke drift problems – a factor blamed as a contributory cause in the Leighton Buzzard accident of 1931 – and in the light of this experience the Southern stuck with the successful side deflectors and equipped all the remaining members of both classes.

During this period the four-cylinder Lord Nelsons were making their appearance and being run in and tested in the expectation that they would be able to haul 500 ton trains at 55mph start to stop schedules. Whether because of their

783 *Sir Gillemere* was equipped with a curious shovel-shaped mounting around the chimney in an early attempt to solve the smoke drift problem, April 1927. (ColourRail)

*772 **Sir Percivale***, with the successful but ugly German smoke deflectors, passes
Bromley with a continental express for Dover, just two months before entering
Eastleigh Works for overhaul, when its experimental smoke deflectors were
replaced by the standard Southern Railway design, 1 August 1932.
(J.M. Bentley Collection/H.C. Casserley)

initial disappointing performance
or limits on the Chief Mechanical
Engineer's investment budget as
electrification became a priority,
authorisation was limited to just
sixteen locomotives and thus they
were never in a position to take
over all the key expresses from
the King Arthurs. Loads of the
fastest expresses therefore stayed
within the capacity of the N15s
and they continued to dominate
the Salisbury–Exeter route and
share on an equal footing the
Waterloo–Salisbury and the
Victoria–Dover expresses with
the new and theoretically much
more powerful 4-cylinder engines.
In Cecil J. Allen's Locomotive
Practice and Performance articles
in the 1929 *Railway Magazine* he
quotes a correspondent seeing
851 coming up from Dover
with 440 tons 'making a terrific
noise' whilst another was almost
noiseless. Talking to the two drivers
concerned, he established that the
first was driving the engine in the
traditional South Eastern fashion,
with one third regulator and 38%
cut-off, whilst the second was full
regulator and adjusting the cut-off
for the gradient. The first driver
stopped at Tonbridge and took
a full five minutes to get away,
with the driver furiously saying
the Nelsons were 'no good' and
wanting his Arthurs back!

Performance of both Urie and

766 *Sir Geraint* in Bulleid's malachite green livery after overhaul and repainting at Eastleigh in March 1939, seen here running in on a local service at Bournemouth Central, before returning to its home depot, Stewarts Lane. (J.M. Bentley Collection)

The classic pose of King Arthur 768 *Sir Balin* at Stewarts Lane fitted with the standard Southern Railway standard smoke deflector design that was eventually fitted to all the 2-cylinder 4-6-0s, c1935. 768 had been fitted with a Lord Nelson eight-wheel flush sided tender for a stint during the early 1930s over the Salisbury – Exeter route, where it gained an excellent reputation, before returning to the Eastern Section and regaining a bogie tender in March 1932. (MLS Collection)

30789 Sir Guy in recently applied sage green livery, at Templecombe, March 1939. (ColourRail)

Maunsell N15s west of Salisbury was particularly good during the later 1920s, whilst east of Salisbury and on the Kent Coast lines there were still problems. The Urie N15s, despite their unaltered valve gear, showed an excellent turn of speed, with 747 *Elaine* recording the first authenticated 90mph on a scheduled passenger train on SR metals (at Seaton Junction descending Honiton Bank on the up *Atlantic Coast Express*). The Scotchmen were stronger on the banks however, capable of sustaining speeds in the upper twenties on Honiton Bank with train loads in excess of 400 tons, around 5mph or more faster than their Urie sisters. In 1928, with increasing holiday traffic to Devon and Cornwall, West of

England trains were permitted to take fourteen coaches of the new Maunsell rolling stock weighing 450 tons tare, but after a season this was adjudged too severe for the N15s without a relaxation of the fastest schedules. On Honiton Bank with 14-coach loads, the King Arthurs were being flogged with full regulator and 55-65% cut-off, whilst the Urie N15s were flat out requiring 70% cut-off if the boiler could maintain steam. One recorded run with a Urie N15, 755 *The Red Knight* on a 480 ton gross load, took 105 minutes net to get from Salisbury to Exeter, falling to 18mph on Honiton Bank and losing eight minutes on the normal schedule, although working timings were apparently adjusted without altering the public timetable,

a deplorable practice that was prevalent on BR's Western Region in the late 1950s.

However, the build-up of traffic, particularly in the summer over the West of England route, caused Maunsell to think of other motive power solutions. In 1933 he had drawings prepared for a massive 4-6-2 + 2-6-4 Beyer Garratt intended initially for continental expresses on the Eastern Section, but when that proposal was rejected for weight reasons, a serious consideration was given to ten such locomotives, to be manufactured by Beyer Peacock, for exclusive operation between Salisbury and Exeter, replacing twenty-three Maunsell King Arthurs, six Urie N15s, sixteen S15s and nine H15s. The proposed locomotive was to have six cylinders, 6ft 3in diameter coupled wheels, grate area of 51.6sqft, and working boiler pressure of 220psi, with a coal capacity in the bunkers of seven tons and 6,000 gallons of water. This huge engine would weigh well over 200 tons, a 25% increase over the weight and size of the LNER and LMS Beyer Garratts, which were, of course, purely freight engines. Despite a financial case being made out for these locomotives around 1935, little more was heard and the project appears to have been dropped. The King Arthurs continued to reign supreme west of Salisbury.

As well as restriction of funds for any new construction of steam locomotives, the possible reason for the demise of the Garratt proposal, the maintenance side was also starved of investment and although Eastleigh Works was relatively modern, the maintenance

In the Second World War, 783 *Sir Gillemere* was equipped with a curious pair of chimneys to attempt to confuse enemy aircraft by laying a smokescreen to camouflage its train. The experiment was unsuccessful and a standard chimney was restored within a couple of months, January 1941. (MLS Collection)

sheds were old and dirty, and having to maintain the increasing number of express engines that were now being more heavily utilised than ever. Loads were also increasing, so that the fleet of both Urie and Maunsell King Arthurs entered a new decade with potential problems ahead, and, faced with increasing competition, the Southern general management had decided to accelerate their main expresses on the most important trunk routes – to Dover, Bournemouth and Exeter. How they fared will be described in chapter 9.

During the Battle of Britain and threat of attacks over Kent, attempts were made to make the passenger services less vulnerable by using the locomotive to lay a smokescreen over the train as the naval vessels also did. 783 was selected for trials, and in November 1940 it was given three small diameter stovepipe chimneys – this had the desired effect on the smoke dispersal but unfortunately destroyed the engine's capacity to steam properly. A month later two larger diameter stovepipes were tried which improved performance but obscured vision of the driver rather than the enemy pilots, which was self-defeating.

In the meantime, unlike its Exmouth Junction companion knights, 792 *Sir Hervis de Revel* had acquired a poor reputation and Bulleid equipped it with a Lemaître multiple jet blast pipe and large diameter chimney to improve its steaming. But even this was to no avail and it was not until the standard Maunsell King Arthur chimney was restored in 1952 that, for some inexplicable reason, all was put right and 792 was able to perform at the same level as other members of the class.

During the war, the malachite green livery had been replaced by plain black, but from 1946 engines were once more being painted in

792 *Sir Hervis de Revel* was the 'black sheep' of the Exmouth Junction allocation of Scotchmen and it was fitted with a wide diameter chimney and multiple jet exhaust in 1940 in an effort to improve its steaming – which apparently failed to improve its reputation. Seen here at Eastleigh in the immediate post-nationalisation period, before renumbering, c1949.
(J.M. Bentley Collection)

the malachite green livery – this even lasted into 1949 when the British Railways numbering and lettering would be imposed on the former Southern Railway livery. No modifications to the basic King Arthur design were made in the BR era, apart from the replacement of the Drummond 'watercart' tenders and some of the six-wheel tenders by the 5,000 gallon bogie tenders, when the Eastern Section King Arthurs were transferred to the South Western route after the Kent Coast electrification in 1959. All the King Arthurs were painted in the standard British Railways passenger brunswick green livery from 1949, with the small lion logo on the tender being replaced by a modified design from 1957. The first withdrawals took place at the end of 1958 and the class was extinct with the condemning of the

30786 *Sir Lionel* outshopped in malachite green with British Railways numbering but awaiting a decision on the style of BR logo or lettering for the tender, July 1949. This was the last King Arthur to be painted in this livery before the application of the BR brunswick green. (John Scott-Morgan Collection)

last survivor, 30770 *Sir Prianius*, from Eastleigh depot in November 1962.

There had been a shortage of coal after the freezing winter of 1947-8 and the best quality steam coal was being exported to earn Britain valuable foreign currency, so the coal available for British Railways' engines on all Regions suffered, and performance in the first couple of years of nationalisation had scarcely recovered from the war years. The Southern Region was no exception to this and the result was many runs with locomotives experiencing steaming problems and also throwing fire, the burning coal dust exhausted starting numerous lineside fires. In an attempt to cure the latter, two King Arthurs were fitted with spark arresters contained in large diameter chimneys which resembled the shape of Beefeaters' hats, thus earning this nickname for the engines so converted, 784 and 788. Whilst it had some impact on the

fire throwing it unfortunately had even more effect – in a negative way – on the engines' ability to steam, especially 784. 784's chimney was removed in May 1948 and replaced with a new design, similarly shaped, in February 1949. It improved the steaming but was not particularly effective in

stopping the emission of sparks. 788 lost its spark arrester in June 1951 and 784 soldiered on until October 1954, when a conventional King Arthur chimney was refitted. By this time coal availability of a decent quality was assured.

30456 *Sir Galahad* with BR number and lettering but still in malachite green livery at Andover Junction on a down Waterloo – West of England train, June 1949. (R.K. Blencowe Collection)

30786 *Sir Lionel* in malachite green livery, renumbered but still showing 'Southern' on the tender, March 1949. (ColourRail)

30784 *Sir Nerovens* fitted with the spark arrestor 'Beefeater' chimney between 1949 and 1954. It is photographed here at Bournemouth Central shortly before its Works overhaul when it received the standard 'Arthur' chimney, 28 August 1954. (Colin Boocock)

30449 *Sir Torre*, refreshed after its last general overhaul at Eastleigh, 7 September 1957. It acquired a 5,000 gallon bogie tender from the withdrawn 30753 in January 1958, and was withdrawn in December 1959, having worked 1,373,426 miles. (John Hodge)

A good view of the Drummond watercart tender on 30451 *Sir Lamorak* on Nine Elms shed, c1956. It received a bogie tender from withdrawn N15X 32333 in January 1957.
(MLS Collection)

One of the three Nine Elms Eastleigh Arthurs in the mid-1950s, 30456 *Sir Galahad*, still with 'watercart' tender, on Basingstoke shed, c1957. It did not receive a bogie tender (from 30749) until August 1958.
(MLS Collection/J. Davenport)

30774 *Sir Gaheris* at Stewarts Lane before its transfer to Nine Elms in July 1955.
(MLS Collection/J. Davenport)

30805 *Sir Constantine*, allocated to Dover after its general repair in June 1957. 30805 was transferred to the Western Section of the Southern Region in June 1959, allocated to Eastleigh, but did not receive a bogie tender before its withdrawal in April 1961. Photo c1957.
(MLS Collection)

30457 *Sir Bedivere,* ex-works at Eastleigh before returning to its home depot of Nine Elms, March 1959. It is uniquely attached to a high-sided 5,200 gallon bogie tender which it received from H15 30490 when that engine was withdrawn in 1955. 30457 was withdrawn in May 1961 having completed 1,429,723 miles in traffic. (J.M. Bentley Collection)

A portrait of 30799 *Sir Ironside* (alias *The Red Knight*) with six-wheel tender at about the time it was transferred from the Southern Region's Eastern Section to Salisbury depot, after the Kent Coast electrification, June 1959. 30799 retained this tender until withdrawal in February 1961. (P.H. Groom)

30774 *Sir Gaheris* receiving its last heavy overhaul at Eastleigh Works in 1957. It was withdrawn in January 1960, having run 1.12 million miles, 7 September 1957.
(John Hodge)

30786 *Sir Lionel* stopped for casual repairs at its home depot, Eastleigh, alongside BR Standard 2-6-4T 80010, 7 September 1957.
(John Hodge)

A work-stained Salisbury King Arthur, 30450 *Sir Kay*, at Nine Elms, by then matched with a 5,000 gallon bogie tender from withdrawn Urie N15 30737, shortly before withdrawal in September 1960, having run 1,478,783 miles.
(P.H. Groom)

30770 *Sir Prianius* undergoes a 'Heavy General' overhaul at Eastleigh Works for the last time on 25 August 1957. It subsequently received a number of 'Light Intermediate and Casual' repairs at the Works before its withdrawal as the last active King Arthur in November 1962, having worked 1,144,608 miles in traffic.
(R.C. Riley)

Six-wheel tender Arthur, 30793 *Sir Ontzlake,* at Eastleigh after a 'Heavy General' overhaul and repaint, 25 August 1957. It was transferred to Feltham in June 1959 and withdrawn from Basingstoke in September 1962, having received a Urie bogie tender for use on the Southern Region's Western Section. The busy scene at Eastleigh includes Merchant Navy 35026 *Lamport & Holt Line* and M7 30029 in the front line of engines for Eastleigh Works attention and BR Standard 4 2-6-0 and Standard 3 2-6-2Ts in the background together with 34006 *Bude,* distinguishable by its extended smoke deflector. (R.C. Riley)

30802 *Sir Durnore* at Stewarts Lane prior to working a special relief train, 30 March 1959. Already equipped with a bogie tender, acquired from the withdrawn 30750 in June 1958, it would be transferred to the Western Section at Eastleigh in June 1959 and be withdrawn in June 1961. (R.C. Riley)

8.2 S15s, design, construction and early operation

Despite the Southern Railway Board's policy of electrification, steam traction remained the only source of motive power for the freight business. Third-rail electrification was not suitable for goods yard working for staff safety reasons and therefore Maunsell needed to consider the future goods requirements as well as the need for a larger pool of efficient passenger engines. Although most heavy industry was outside the boundaries of the Company, about 25% of its revenue came from freight, including coal from Kent and that imported from South Wales via Salisbury and the West Midlands via Oxford and Basingstoke. There was extensive agricultural traffic from the West Country and seasonal traffic from the Channel Islands, the Isle of Wight, Hampshire and Kent, but the most regular traffic was the flow of general merchandise for shipment to and from Southampton Docks, which were still being further developed by the Southern Railway. Much of this freight flowed to and from the other railways, some via Basingstoke and Oxford, but most via London through Feltham Yard and the cross-London lines to Acton, Willesden, Brent and Ferme Park. The Southern Railway had its own London Goods depots at Nine Elms and Bricklayers Arms. Feltham was the Company's largest marshalling yard, but Norwood was the focus for traffic to and from the former LB&SCR and Hither Green the South Eastern routes.

As mentioned in chapter 6, Maunsell had carried out comparative tests with the freight engine constituents of the three railways that formed the Southern Railway, and had found the Urie S15 4-6-0s to be the best and most

Maunsell S15 836 as built in December 1927 and fitted with a flush-sided eight-wheeled tender, which it later exchanged for a King Arthur bogie tender before again receiving a six-wheeled tender to work on the London – Brighton route.
(John Scott-Morgan Collection/F. Moore)

Maunsell S15, 833, built with eight-wheel tender in November 1927, at Feltham after swapping tenders with a Lord Nelson and gaining a bogie tender from King Arthur 764, on entry into traffic, May 1928. (J.M. Bentley Collection/Loco Publishing Co.)

833 after receiving smoke deflectors in July 1930, at its home, Feltham depot, returning from a General Overhaul in November 1930. (J.M. Bentley Collection)

suited of the classes for the traffics on offer, especially that between London and Southampton Docks. Maunsell had no hesitation, therefore, when additional freight engines were deemed necessary, in applying the same principles in modifying the Urie design as he had improved the N15 class. He obtained the Southern Railway Board's authority to build ten S15s, 823-832, at Eastleigh Works at a cost of £6,585 each (Bradley in *LSWR Locos*; the RCTS book quotes £10,415 – possibly the cost of the bogie tender not included in the first quotation). These engines were constructed between March and November 1927 and were followed immediately by 833-837 between November 1927 and January 1928.

The new goods engines were fitted with two 20½in diameter x 28in stroke cylinders, 11in piston valves with long travel, outside steam pipes and 200psi boiler pressure, all dimensions similar to his King Arthurs. The boilers were fully interchangeable with other classes carrying the N15 pattern. They also had the improved draughting, enlarged ashpans and Maunsell superheater. The driving wheels were 5ft 7in, the same as the Urie engines, but the cab outline and straight running plate again followed the appearance of the Scotch Arthurs. The grate area at 28sqft was slightly smaller than the N15s, but the weight was almost identical at 80 tons 14cwt for the locomotive and 137 tons 2cwt including a heavy bogie tender, the engines from 833 having straight

S15 836, built in January 1928, now equipped with a six-wheel tender ex-King Arthur 766, so that it could be allocated to the former LB&SCR lines, in SR green passenger livery, July 1937. (J.M. Bentley Collection)

flush sides instead of the 'lipped' rims. Tractive effort was 29,860lb. In the late 1920s the Maunsell and Urie 4-6-0s were all suffering from drifting smoke problems, and when a standard smoke deflector had been eventually approved after a series of experiments on the King Arthurs, the Maunsell S15s all received smoke deflectors between 1929 (just 831) and 1931.

Initially the 823-832 batch of engines were equipped with 5,000 gallon bogie tenders with flared rims, similar to the Urie S15s and the first series of Scotchmen. However, in 1928, a confusing set of tender exchanges took place affecting all the Southern 4-6-0s, and the S15s in particular. The first exchanges were made because of the need for the 1926 Lord Nelsons to have larger tenders and 833-837 gave up their straight-sided bogie tenders to 851 and 854-857. In return, they received 5,000 gallon bogie tenders with the flared tops from N15s 763/64/68/71/72 when the latter received six-wheel tenders sufficient for Eastern Section working. Then some tenders from the 823-832 series were transferred to Lord Nelsons 852-853 and 858-860, and in return S15s 828-832 received flared top bogie tenders from Eastern Section N15s 765-767, and 769-770. At the last moment, 769 was transferred to Exmouth Junction and retained its 5,000 gallon tender so 831 held on to its own tender. Therefore 852 *Sir Walter Raleigh* had to run with a smaller tender for a while. Then, because S15s 833-837 were required for goods traffic on the Central ex-LB&SCR Section, smaller wheelbase tenders were needed to fit the turntables there as with the 793-806 series of N15s, and they obtained six-wheel tenders from Eastern Section N15s 763-767 as the latter engines were frequently loaned to Nine Elms, where the larger bogie tender was essential.

Initially, 823-837 entered traffic in goods lined black livery, but as they frequently appeared on passenger turns, being considered in effect 'mixed traffic' locomotives, the 838-847 series was painted in the green passenger livery and the earlier engines were repainted similarly in the mid-1930s. After running in from Eastleigh depot on freight services to Salisbury or stopping passenger trains to Bournemouth, 823-827 were allocated to Exmouth Junction, 828-832 to Salisbury and 833-837 to Feltham. They were immediately well received by the crews who found they were free-running engines with a better turn of speed than the larger wheeled H15s, and they were therefore frequently used on passenger turns, especially west of Salisbury. The Urie S15s could achieve 60-65mph on occasions, but the new Maunsell engines soon demonstrated that running at 70mph was no problem. Before its electrification, the Feltham engines often found themselves on passenger services on the direct Portsmouth line, where they acquitted themselves well and were the equal of the Schools class, with their greater power for

1936-built 840 with eight-wheel flush-sided tender seen here shortly after the Second World War in wartime plain black livery at Feltham Yard, c1946. (MLS Collection)

hill climbing compensating for the lower speeds on the level and downhill sections.

A final batch of ten, 838-847, was ordered in March 1931, but because of the economic crisis of the Depression, these locomotives were not in fact built until the recovery in 1936, when costs had reduced and government assistance was available. The output from the Kent coal mines had grown rapidly and Maunsell sought to provide something for the traffic department more powerful than the 'N' class Moguls, which were still the prime freight power on the former SE&CR routes. Around 1933-4 he drew up plans for a heavy freight 4-8-0 with four cylinders, a large boiler similar to the Lord

Salisbury S15 30829, as built with 5,000 gallon bogie tender, ex-works at Eastleigh in March 1955. (J.M. Bentley Collection)

Nelsons though at a lower pressure of 200lb, 5ft 1in coupled wheels and 33sqft grate area. The engine would be provided with a 5,000 gallon flush-sided tender, similar to that of the Nelsons, giving a total weight in running order of over 144 tons. However, he ran into problems with the Civil Engineer again, and although he struggled and managed to reduce the maximum axle weight to 21½ tons, substantial investment was needed to improve the track and lengthen loops to take advantage of its increased pulling power. The lack of money available when the Southern's priorities were on electrification meant that the design was never developed to the construction phase. The outline drawing of this proposed design is in Appendix 8. Another proposal for a mixed traffic 2-6-2 with 6ft 3in coupled wheels and weighing 148 tons foundered for the same reasons. In the end, it was decided that a new batch of S15s, previously authorised in 1931, would be better value and would meet the need on the Kent coalfield as well as improving the transit of fast freight traffic to the channel ports.

When this final batch of ten S15s was built in 1936, 843-846 were allocated to Feltham, enabling 833-837 to be transferred to Brighton.

30829, still based at Salisbury, awaiting Works attention at Eastleigh in company with a G6 0-6-0T and a Lord Nelson, May 1959. After overhaul it would run for a further four and a half years, being withdrawn in November 1963, after working 1,209,387 miles in traffic. (Colin Boocock)

Another shot of 30833, with six-wheel tender, at Eastleigh after receiving an 'Intermediate' overhaul (only the smokebox repainted), February 1959, sandwiched between a T9 4-4-0 and an ex-LMS 2MT 2-6-2T. (Colin Boocock)

Salisbury's 1927-built S15, 30824, pauses between shunting at Broad Clyst, near Exeter, 6 July 1961. (R.C. Riley)

847 went new to Exmouth Junction and 838-842 were allocated to Hither Green. The Central Section engines replaced the Billington 'K' Moguls and had four regular freight diagrams (turns 695-699), mainly between Battersea Yard or Norwood to Brighton or Hove. They were significant improvements on the previous Moguls, having more power, stronger acceleration and better braking. For some reason they were restricted to 45mph and prohibited from working passenger or empty stock trains in contrast to the use of their sisters on the Western Section. The five Hither Green 4-6-0s took over freight services to Ashford and Dover and some coal traffic from the 'N' class Moguls – they were also restricted to 45mph on this Section. The Salisbury and Exmouth Junction engines worked many semi-fast passenger services between Salisbury and Exeter as

well as all freight traffic emanating from the West Country. On summer Saturdays, the Maunsell S15s regularly deputised for the Urie and Maunsell N15s on relief services to the West of England, Bournemouth and boat trains to Southampton Docks.

8.3: N15X, design, construction and early operation

After the building of the Marsh 'Atlantics' between 1904 and 1906, the LB&SCR management decided that tank engines would be sufficient to run all their express passenger traffic. The precursors of this policy were the very successful superheated 'I3' 4-4-2 tank engines which lasted working Victoria–Oxted passenger trains until replaced by Brighton-built LMS Fairburn 2-6-4 tanks in the early 1950s. However, their water

capacity was stretched to the limit if they were delayed on the longer distance services and something larger was needed. In 1910 Marsh built a 4-6-2 tank, 325 *Abergavenny*, and a second one was built by his successor, L.B. Billington, to a similar design, though with Walschaerts instead of Stephenson valve gear.

Although both the I3s and the Pacific tanks performed admirably, Billington sought something even larger, and in April 1914 the first 'Baltic' 4-6-4 tank entered service, numbered 327. No.328 followed and then there was a long interruption because of the First World War, with 329-331 being built in 1921 and 332-333 following in 1922 just before the Grouping. It had been intended to complete ten of these locomotives but, overtaken by the formation of the new Southern Railway, 334-336 were never built. They had 6ft 9in coupled wheels, two 22in x 28in cylinders, and boiler pressure of 170lb. Their tractive effort was 23,325lb. With around 2,700 gallons water capacity in their tanks, they weighed just over 98 tons. They were designed to maintain the fast 60-minute schedules of the London–Brighton non-stop services and performed admirably, being fast runners and very smooth riding. The last engine to be constructed, 333, became the Company's war memorial engine, being named *Remembrance*.

Their reign on the Brighton and Worthing trains came to an end on 1 January 1933 when the Southern Railway's electrification project on the Central Section was inaugurated. 2332 (the seven Baltic tanks were renumbered 2327-2333

333 *Remembrance*, the LB&SCR 'Baltic' 4-6-4T, that was converted to N15X 2333 in 1935, at Brighton, c1922.
(John Scott-Morgan Collection/ Real Photographs)

B331 (unnamed), that was rebuilt as N15X 2331 in 1936 and named *Beattie*, photographed here in its last year before rebuilding, 1935.
(John Scott-Morgan)

Baltic Tank B332 with a Brighton line express near Purley, c1923. (Colin Garratt Collection/Rev A.W. Mace)

by the SR) hauled the last steam *City Limited* from London Bridge at 5pm on 31 December 1932 and reached Brighton in 57¾ minutes with its heavy 385 ton load. Earlier, 2333 had worked the last down *Southern Belle* Pullman train reaching a maximum of 77mph

before signal checks spoiled the run into Brighton, though it was still reached in 58¾ minutes from Victoria.

After being replaced on the main Brighton route, the Baltic tanks worked from Eastbourne until that route was electrified in July 1935.

The Pacific tanks remained there to work trains like the *Sunny South Express* between the Midlands and the South Coast, but there was little work left of substance for the Baltic tanks.

By the mid-1930s Maunsell was being frustrated in his efforts to provide more powerful engines for the Eastern and South Western main lines through lack of investment cash, but the redundancy of these large tank engines on the routes for which they were designed gave Maunsell the opportunity to rebuild them into tender engines for use elsewhere. A number of modifications were made to the dimensions as well as the obvious ones in converting to a tender engine. The boiler pressure was increased to 180psi and the large cylinders were lined up to 21in x 28in. The rebuilt engines weighed 73 tons and the Urie 5,000 gallon tenders weighed 57 tons 11cwt. The rebuilt engines were all given the names of historic locomotive engineers, apparently suggested to Maunsell by his assistant, Harry

Baltic Tank 2328, in its last year before rebuilding as a 4-6-0, at Polegate with a Victoria–Eastbourne route express, 1935. (Colin Garratt Collection/Rev A.W. Mace)

Holcroft, although 2333 retained its LB&SCR war memorial purpose and name. They were:

2327 *Trevithick* (formerly Charles C. Macrae, the name of a previous LB&SCR Chairman)

2328 *Hackworth*

2329 *Stephenson* (this engine had been named in 1921 after the last survivor of the Stroudley single-wheelers)

2330 *Cudworth*

2331 *Beattie*

2332 *Stroudley*

2333 *Remembrance*

The first engine to be rebuilt at Eastleigh Works in December 1934 was 2329, followed by 2327, 2330, 2332 and 2333 in 1935 and, finally, 2328 and 2331 in February and May 1936 respectively. They received 5,000 gallon tenders from Urie S15s which in turn got tenders from Drummond's withdrawn C8 4-4-0s. Although not a rebuild of the King Arthur family, they were sufficient in power and purpose to be classified as N15X, and the hope was that the excellent performances of the Baltic tank originals would be maintained in their new form.

Initially they were allocated to Nine Elms where their work was mainly on the Bournemouth line. At first they were entrusted with prestige trains such as the *Bournemouth Belle* but even though that train was limited to ten Pullman cars at that time, the rebuilt engines were disappointments. They did not have the advantage of a modern front end, and could not really be expected to compare with Maunsell's King Arthurs, although they should have been comparable to the Urie N15s. They were soon relegated to semi-fast and secondary expresses, and summer reliefs on the Bournemouth route, and made a useful addition to the South Western Section locomotive fleet, although they never lived up to the reputation that they'd gained on the Brighton line in their original condition.

2329 *Stephenson* after rebuilding at Eastleigh from LB&SCR Baltic Tank, c1935. (MLS Collection)

Newly rebuilt N15X
2333 *Remembrance*, at
Waterloo, 1935.
(John Scott-Morgan Collection/
Photomatic)

32329 *Stephenson* in
malachite green, but BR
number on cabside and
'Southern' on the tender,
Eastleigh, c1948.
(MLS Collection)

32329 *Stephenson* backing onto its train at Basingstoke, c1955. (R.C. Riley)

32333 *Remembrance* in BR mixed traffic lined black livery, at Basingstoke station, taking water, c1955. (MLS Collection/W. Potter)

32329 *Stephenson* ex-works at Eastleigh, c1955.
(MLS Collection)

32328 *Hackworth* in BR mixed traffic livery at Bournemouth station, a T9 in the background, c1955. (Colin Boocock)

The last surviving N15X, 32331 *Beattie*, at Nine Elms, c1957.
(MLS Collection/P.H. Groom)

A portrait of 32329 *Stephenson* in the BR mixed traffic livery of lined black at Eastleigh, 1956. (R.C. Riley)

Another view of 32329 *Stephenson* at Eastleigh, 1956. (R.C. Riley)

32331 *Beattie* at its home depot, Basingstoke, in its last year of service, 1957. (R.C. Riley)

Chapter 9

URIE AND MAUNSELL ENGINES IN SOUTHERN RAILWAY OPERATIONS

9.1: Urie's H15s

Before the Grouping, H15s were rarely seen on the Portsmouth direct route, apart from occasional bridge testing in the Fratton area. However, shortly afterwards they began to appear regularly on Portsmouth expresses, along with Drummond D15s, T14s and Urie N15s and S15s. They were especially useful in the summer with heavy holiday crowds making their way to Portsmouth Harbour for the Isle of Wight. In the early days of the Southern Railway, after the building of the King Arthurs, they were mainly employed on mixed traffic duties – semi-fast passenger services to Basingstoke and Salisbury, freights to Southampton Docks and the occasional Ocean Liner express.

In the winter of 1924-5, O.S. Nock was a frequent traveller to and from Bournemouth and obviously was keen to experience the performance of the new H15s which were initially allocated to Eastleigh. He expressed himself disappointed at the time, and recorded a typical run on the 6.30pm Waterloo-Bournemouth that winter, when

Newly-built Urie H15 488 draws a train of L&SWR coaches out of sidings at Eastleigh, c1914. (MLS Collection/G. Goltas)

The Urie rebuilt E14, now H15 335, arrives at Chard Junction with a stopping train for Exeter, c1920. (J.M. Bentley Collection)

523 had an eleven-coach 336 tons tare train, loaded to 355 tons. Signal checks out of Waterloo as far as Surbiton cost 6-7 minutes, but the train fell from 60mph passing Weybridge to 42½ mph at milepost 31 and just managed 60 on the level before Basingstoke, falling to 44mph at Worting Junction and reaching a maximum of 70mph

through Winchester, before arriving nearly 6 minutes late on the easy 92 minute schedule to Southampton Central. With a maximum of 65mph down Hinton Admiral bank, and a permanent way slack after Christchurch, another three and a half minutes were lost to Bournemouth.

In 1925, the maximum loading for the continental expresses from Victoria was raised from 300 to 425 tons in anticipation of the new King Arthurs replacing the D1 and E1 4-4-0s. Unfortunately, the new Scotchmen had a chequered start to their career, with some flaws in the boiler construction at the North British Works causing the engines to be short of steam on many occasions. Time was being lost, especially on the banks between Orpington and Tonbridge, and the new Lord Nelson class was not yet ready. Whilst Maunsell sought a cure for the deficiencies of the N15s,

Nine Elms H15 483 standing at Basingstoke with a Salisbury–Waterloo semi-fast train, c1930. (ColourRail)

he decided to try out one of the new H15s which he had just had constructed at Eastleigh to Urie's basic design, though with the N15 boiler.

478 was stationed at Bricklayers Arms in February 1927 and was put to work on the Victoria–Dover boat trains. However, performance on the banks was even worse, just as Eastleigh and Bournemouth crews were experiencing on the climb to Litchfield summit, and it was soon transferred back to the Western Section. It was curious that at this time, performance of the 1924 engines was inferior to the original Urie 1914 ones, despite the use of the N15 boiler, which was a success when carried by 491 of the original series. 483, 490 and 522 were considered to be the best of the class at this time, with 521 and 524 being the 'black sheep'.

330, 331 and 333 were transferred

H15 330 with a down West of England express reaches the summit at Hewish Gates Box, 2 August 1928. (H.C. Casserley)

Rebuilt F13, No.333, at Axminster with a stopping train to Exeter, 1927. (Colin Garratt Collection/Rev A.W. Mace)

from Salisbury to Nine Elms in June 1935 to cover additional ocean liner specials and Bournemouth excursions, but the London crews did not cope with them as well as the Salisbury men, and they swiftly returned to their home base after the summer timetable. The most remarkable performance of any of the Drummond rebuilds that I have seen was recorded by O.S. Nock sometime in the 1930s – he never put a date on it. It was exceptional to get an H15, especially one of the rebuilds, on the *Atlantic Coast Express* in the first place, and he was amazed at the quality of the run that ensued, so remarkable that it is worth tabling in full. Nock joined the train at Exeter Central when the load was a manageable ten coaches, but three additional well-filled coaches were added at Sidmouth Junction, making an unusually heavy load for this schedule. Most noteworthy were the high speeds he tabled with this engine – the H15s were more known for their hill-climbing ability than free running downhill.

From the 1935 winter timetable the allocation and main turns of the twenty-six H15s were as follows:

Nine Elms: 473, 474, 482-491: there were two weekday diagrams covering Waterloo–Salisbury semi-fast passenger trains during the day and Nine Elms–Salisbury fast freights at night; a freight from Nine Elms to Fratton, then passenger to Eastleigh returning to London with another goods; two diagrams covering a Nine Elms–Dorchester goods, followed by Weymouth–Bournemouth stopping trains, returning to London on a freight from Dorchester; a Waterloo–Southampton Docks van

Atlantic Coast Express – Sidmouth Junction–Salisbury 333 – H15 (Rebuilt Drummond F13) 13 coaches, 419/450 tons			
Location	Actual time	Speed (mph)	Mileage
Sidmouth Junction	00.00		
MP 158	03.13	45	1.3
Honiton	09.44	24½ / 26½	4.6
MP 153½	12.35	22½	5.8
Seaton Junction	18.34	83½	11.5
Axminster	21.07	71	14.8
Chard Junction	26.18	54	19.9
MP 133½	34.07	37½	25.8
Crewkerne	36.38	75	27.9
MP 126½	41.26	52	33.1
Yeovil Junction	45.03	69	36.7
Sherborne	49.28	57	41.3
MP 115½	53.15	28	43.9
Templecombe	58.26	68	47.4
MP 107½	63.31	39½	51.9
Gillingham	66.05	54½	54.2
Semley	72.16	29½	58.3
Dinton	81.40	74	67.6
Wilton	86.56		73.3
Salisbury	90.18		75.8

train returning on a goods; and finally a Nine Elms–Salisbury fast freight and return.

Eastleigh: 476-478, 521-524: these locomotives only had one regular diagram, a goods from Eastleigh to Salisbury and return. However, they were mainly 'on call' for passenger, van and goods specials from Southampton Docks to London.

Salisbury: 330-335, 475: these engines had four regular weekday diagrams, the first being local passenger services to Eastleigh and Basingstoke during the day and a Salisbury–Nine Elms goods at night. Another worked Salisbury–Waterloo semi-fasts and a Salisbury–Basingstoke goods train. A third was rostered to a Salisbury–

Southampton freight and the last worked a Salisbury–Exmouth Junction goods and returned from Sidmouth Junction to Salisbury with a loaded milk train.

There were additional turns on summer Saturdays booked for N15s, but H15s would often be substituted if the passenger engines were not available.

There was one run recorded in Cecil J. Allen's 'British Locomotive Practice and Performance' in a late '30s *Railway Magazine* that reached the highest speed I've seen in print with one of the H15s built during the Maunsell era. Salisbury's lone 475 was rostered one day to the 1pm Waterloo–Exeter between London and Salisbury, stopping at Woking and Andover, normally a

Maunsell-constructed H15 to Urie design with tapered boiler, 473, passing Clapham Junction with a Waterloo – Portsmouth train, May 1931. (J.M. Bentley Collection/Photomatic)

turn for a Salisbury King Arthur. The load was a comparatively light one, but performance was well above normal for this series of engines (see page 107).

When Bulleid became Chief Mechanical Engineer in 1937, he conducted a number of experimental livery changes before settling on malachite green for the main passenger classes. 330, 473, 475, 486 and 524 remained in Maunsell green with black and white lining, but with Bulleid lettering and cabside numerals. 332 and 491 appeared in olive green with green and yellow lining, 477, 482, 485 and 488 in olive green with black and yellow lining. 334, 335, 478, 483, 487,490 and 523 had plain Maunsell unlined green, and 474, 476 and 489 sported unlined malachite green. All now had Bulleid lettering and cabside numerals. Then, during the war, all were repainted plain black,

Nine Elms Maunsell H15, 524, passes Esher with a Portsmouth–Waterloo train, September 1931. (J.M. Bentley Collection/Photomatic)

An Eastleigh Works scene, with a Urie H15 on the left, taper-boilered H15 475 in the centre and 'Scotch Arthur' 786 *Sir Lionel* on the right, c1930.
(Colin Garratt Collection/Rev A.W. Mace)

Urie H15 490 enters Guildford with a Waterloo – Portsmouth train. The lines on the left lead to Farnborough and Reading, and the electrified suburban line from Guildford to Waterloo via Effingham Junction curve away to the right, c1935.
(Colin Garratt Collection/Rev A.W. Mace)

1pm Waterloo–Exeter (as far as Andover)
475, 8 coaches, 253 / 270 tons

Location	Mins Secs	Speed	
Waterloo	00.00		
Clapham Junction	06.52	45*	
Wimbledon	11.00	52	
Surbiton	15.58	62½	
Walton	–	67	
Weybridge	22.35	61½	
Byfleet	24.55	68	
Woking	27.50		Sch 29
	00.00		
Brookwood	-	47 / pws	
MP 31	11.47	25	
Farnborough	15.00	52½	
Fleet	18.20	64½	
Winchfield	–	60	
Hook	23.50	67½	
Basingstoke	28.00	62	
Worting Junction	31.35		
Overton	36.52	66½	
Whitchurch	39.55	73½	
Hurstbourne	41.26	77½	
MP 62½	44.35	72½	
Enham Box	–	81½	
Andover Junction	45.58	(43¾ net)	Sch 45

starting with 475, 486 and 521 in March 1941.

Traffic flows changed during the wartime years. Troop trains, van and goods trains to Southampton Docks took precedence and the Nine Elms and Eastleigh engines took these rather than their previous regular diagrams. In late 1940, 333, 484, 486 and 524 moved to Feltham and 521 to Bournemouth for main line freight work. New routes were tested and from August 1941 the H15s were allowed into Sussex as far as Chichester, and Eastleigh and Salisbury engines started to make appearances there. The GWR needed additional freight locomotives and a number of Southern 4-6-0s were loaned including 478, based at Old Oak Common, substituting for an N15X that was in Shops under heavy repair. On return it was used by Basingstoke for a few weeks for Portsmouth services via Eastleigh and troop specials. Other H15s were reported in far-flung places during the war, including Tonbridge, Birmingham and even Chester.

In March 1944 ten new WD 2-8-0s (7436-7445) were delivered to Eastleigh and in consequence 477, 522 and 523 were transferred to Nine Elms, 478 to Feltham and 333, 335 and 476 to Salisbury. More changes took place when the WDs were required in Europe and the H15s were further reallocated to Nine Elms (477 and 482-491), Feltham

(473, 474 and 478), Eastleigh (521 -524) and Salisbury (330-335, 475 and 476). During the build-up for the D-Day invasion, 334, 475, 486 and 491 headed troop trains into Redhill off the line from Reading.

At the end of the hostilities ocean liner specials recommenced, with 486 working the first for the *Queen Mary* in August 1945. Their final Southern Railway allocation in 1947 was:

Eastleigh: 473, 474, 521-524
Nine Elms: 477, 482-491
Salisbury: 330-335, 475, 476
Bournemouth: 478

The latter spent its time working Weymouth line semi-fasts and stopping services until June 1947, when it joined its sisters at Eastleigh. All the H15s entered the service of the nationalised railway in 1948.

9.2: Urie's N15s

After the line speed restrictions of the First World War period and its aftermath, an extensive track relaying programme took place on the L&SWR main lines, especially the Waterloo–Exeter route, and by mid-1922, tighter schedules were introduced with through working of the Urie N15s to Exeter, remanning at Salisbury. The 10am and 11am down trains were timed the fastest, allowed just 204 minutes for the 170 miles including the Salisbury stop. The afternoon return trains were slower, although with additional intermediate stops; the 4.30pm Exeter was allowed 240 minutes and the last up service, the 6.30pm, 260 minutes. Exmouth Junction engines (740-742 and 749) worked the 10.30am and 12.30pm up with the fast timings (206 and

Salisbury's 754 *The Green Knight* climbs to Honiton summit with the up Atlantic Coast Express, 4 August 1928. (H.C. Casserley)

750 *Morgan le Fay* with a substantial down West of England express at Andover Junction, c1925. (Colin Garratt Collection/Rev A.W. Mace)

210 minutes respectively), returning on the 6pm and 9pm Waterloo (the latter a slow overnight train). This through working only lasted for the summer timetable as the N15s were found to be struggling for steam on these long turns, and engine changing at Salisbury was restored just before the Grouping.

Immediately after the Grouping in January 1923, the allocation of the N15s was:

Nine Elms: 736-742, 746-748, 751-755 (753-755 were only delivered during 1923)
Salisbury: 743-745, 750
Exmouth Junction: 749

742 and 747 were loaned to Bournemouth during the summer timetable, and 738 was loaned to Eastleigh for Southampton Docks ocean liner specials. Maunsell then took a significant interest in their performance and instituted tests with the engine with the poorest reputation, 742, as recounted earlier, in chapter 5. As a result Maunsell ordered a number of improvements, although the cost of rebuilding them with a modern front-end and long lap valve gear was, unfortunately, not faced. Performance improved,

Nine Elms Urie Arthur 748, newly named *Vivien*, shuts off steam to obey the 40mph speed restriction round the Clapham Junction curve, with a semi-fast train for Basingstoke and Salisbury, during the General Strike, 25 July 1926. (H.C. Casserley)

753 *Melisande*, now equipped with smoke deflectors, approaches Clapham Junction with a Waterloo – West of England express, 9 April 1928. (H.C. Casserley)

and particularly so west of Salisbury, where the engines could utilise their free running capabilities to charge the following severe gradients. On long level stretches east of Salisbury the continuous demand on the boiler meant the firemen had a harder time to maintain steam pressure.

During the 1926 General Strike, 737 and 739 were converted to oil-burning, as they had been in 1921. They were initially used on priority freight trains to Southampton Docks at night. 739 was then tested on passenger trains to both Salisbury and Bournemouth, with the oil generating about 70% of steam that the same calorific value of coal could produce, although oil-burning 4-4-0, D15 470, was more efficient working Southampton–Waterloo semi-fast services. Once the men were used to these two engines, they were transferred to Nine Elms for express work, including the 11am Waterloo *Atlantic Coast Express* in July 1926, albeit with additional stops with the much reduced passenger service during the strike.

The allocation of the Urie N15s

at the beginning of the summer service in 1928 was:

> Nine Elms: 738, 741, 742, 745, 748, 750, 752, 753 and 755
> Bournemouth: 736, 743
> Salisbury: 739, 749, 751, 754
> Exmouth Junction: 737, 740, 744, 746, 747

The two Bournemouth engines, allegedly poor performers, worked to Oxford on through trains for the West Midlands and North West. At that time the other N15s were used indiscriminately with the Maunsell King Arthurs. In 1929, 737 and 753

were transferred to Eastleigh to work that depot's two Waterloo turns, hitherto H15 diagrams.

The Salisbury and Exmouth Junction engines were popular on the Salisbury–Exeter legs of the main West of England expresses, where they could use their speed to rush the hills and their boilers could feed their larger diameter cylinders in sufficient bursts to power up the gradients. They shared the Salisbury–Exeter diagrams with the Eastleigh Arthurs, 448-457, and Scotchman 768. The Maunsell King Arthurs managed the long level stretches east of Basingstoke or the long gradients on the Bournemouth line where sustained collar-work was necessary. These would drain the Urie engines' boilers and drivers would have to ease their steeds, causing loss of time. D.L. Bradley, in his book, *An Illustrated History of LSWR Locomotives*, quotes an analysis of runs between Salisbury and Exeter in 1929-30 by a friend of his who commuted regularly between the two cities over the two years:

The equal highest was a run with a T9 4-4-0, 717, which also touched

Class	No. of Journeys	Ave. load (tons)	Ave. speed (mph)				
			Semley	Sherborne	Crewkerne	Axminster	Honiton
Urie N15	37	321	50½	79	77½	83	37
King Arthur	31	356	53½	81	79½	85	35½
H15 (330-5)	13	289	45	69½	67	68½	25½
S15 (Maunsell)	19	308	48	71	70½	73½	37

The highest speeds recorded near Axminster were:

N15 (754 *The Green Knight*)	86mph (747 achieved 90mph later)
King Arthur (451 *Sir Lamorak*)	91mph
Scotchman (788 *Sir Urre of the Mount*)	87mph
Six-wheel tender Arthur (800 *Sir Meleaus de Lile*)	89mph (the only one timed here)
S15 (826)	84mph

Up *Atlantic Coast Express*

| | **745** *Tintagel* | | **747** *Elaine* | | |
| | **12 coaches 394/415 tons** | | **11 coaches 330 / 350 tons** | | |
Location	Actual Time	Speed	Actual Time	Speed	Mileage
Exeter	00.00		00.00		0.0
Broad Clyst	08.22	64½	07.30	71½	4.8
Whimple	12.41	31½	11.05	52 /41½	8.5
Sidmouth Junction	18.30	60 (est)	15.50	64½	12.2
Honiton MP153 ½	27.38	29	23.35	31	18.0
Seaton Junction	34.32	pws/71½	29.10	90	23.7
Crewkerne	54.20	40½/77½	45.55	43/80½	40.1
Yeovil Junction	62.10	-	53.25	61/77½	48.9
Templecombe	74.52	32	65.45/70.10	31½/sigs	59.6
Gillingham	81.25	75	78.45	60/42½/68	66.4
Semley	86.31	37	83.45	36½	70.5
Dinton	95.30	70½	92.45	71½	79.8
Salisbury	104.10	(100 net)	101.35	(93 net)	88.0

755 *The Red Knight*
14 coaches 455 / 480 tons

Location	Actual Time	Speed	Mileage	
Salisbury	00.00		0.0	
Wilton	07.40		2.5	
Dinton	15.35	52	8.2	
Semley	27.35	38	17.5	
Gillingham	31.35	76	21.6	
Buckhorn Weston	-	pws		
Templecombe	39.05	58½	28.4	6 Late
Milborne Port	43.05	30	30.8	
Sherborne	46.45	77½	34.5	
Yeovil Junction	50.45	-/46½	39.1	6¾ Late
Crewkerne	59.50	-/28½	47.9	
Axminster	73.55	74	61.0	
Seaton Junction	77.05	-	64.3	
Honiton MP 153 ½	91.20	18½	70.0	
Sidmouth Junction	96.50	76	75.8	12¾ Late
Broad Clyst	102.45	82	83.2	
Exmouth Junction	-	1½ min sigs stand		
Exeter	110.35	(105½ net)	88.0	13½ Late

91mph. At this time the Urie N15s still retained their 22in cylinders. After these were lined up to 21in, drivers on this route reckoned that their hill climbing performance did not have the same sparkle.

O.S. Nock recorded an excellent run with Nine Elms' 745 in its original condition during this period. This is compared with a run timed by Cecil J. Allen with 747 after the Maunsell modifications and brief details of both runs are tabled left.

These two runs demonstrate the hill climbing ability with the heavier load of the engine still with 22in cylinders, compared with the free running of 747, which was the first to record 90mph on the Southern Railway in ordinary passenger service. During the summer of 1928 some trains were allowed fourteen new Maunsell coaches (weighing 455 tons tare) and Nock timed 755 with a gross load of 480 tons one Bank Holiday.

The timing of the above train had been relaxed to allow for the Bank Holiday load, but the train was nearly fifteen minutes late by the public timetable. Another heavy train was recorded from Salisbury in the up direction at about the same time. 751 *Etarre*, with 490 tons gross, laboured to Grateley taking virtually 12 minutes for the first 5½ miles, reaching just 29mph on the four miles of 1 in 169/140, but then touched 75mph at Andover and, after Basingstoke, was between 65 and 78mph all the way to Hampton Court Junction. With a severe signal check costing 5 minutes after Vauxhall, Waterloo was reached in 96 minutes 17 seconds (91½ net). 744 *Maid of Astolat* was renowned as a fast engine at this time (she was a flighty young lady in the Arthurian legends too – the mistress of Sir Lancelot) and was timed on a 350 ton train after Templecombe at 83½ at Sherborne, 85 at Axminster and 82 before the Sidmouth Junction stop. On another down run with 360 tons, 744 surmounted Semley at 45½, touched 83½ at Gillingham, 82 at Templecombe, 82 at Axminster, fell to 19mph at Honiton Tunnel and then charged through Broad Clyst at 88! Exeter was reached, unchecked, in 94 minutes 51 seconds.

The *Railway Magazine* published three logs in 1930 of up trains between Sidmouth Junction and Salisbury, two with Urie N15s, and for comparison, one very unusually with a Lord Nelson.

Both the N15 runs were delayed by p-way slacks between Templecombe and Salisbury, but *Lord Rodney* had an unchecked run, and with 76½ through Dinton, recovered four minutes of the lost time.

However, *Railway Magazine* articles in the 1930s showed that the N15s of both varieties continued to thrive and even improve, and a splendid run on the down *Atlantic Coast Express* was reported in 1933 with a Urie engine at a time when motive power was usually a Maunsell King Arthur.

In the mid-1930s the allocation was:

Nine Elms: 738, 739, 741, 742, 745, 748, 750, 752, 755

Salisbury: 751, 754

Exmouth Junction: 740, 744, 746, 747, 749

Eastleigh: 737, 753

Bournemouth: 736, 743

The annual mileage of the Urie N15s during the 1930s was high. Nine Elms engines averaged 44,500; Salisbury engines 40,000; Eastleigh engines 37,500; and the Exmouth Junction engines, highest of the lot, at just over 45,000. Scheduled mileage between 'Heavy' Works repairs was 70,000, upped to 85,000 in 1936 and, between 1933

and 1939, the N15s averaged 85,269 and the King Arthurs, 84,546. Both classes achieved this mileage in 21-22 months between main Works overhauls. The H15s and N15X class achieved similar mileages between heavy repairs

although they took 26-27 months to achieve these because of their lower annual mileages. Most of these locomotives, however, were receiving an Intermediate Works repair at around 52-54,000 miles so, although creditable, they did

Sidmouth Junction–Templecombe

Location	740 *Merlin* 230 / 240 tons			745 *Tintagel* 341 / 360 tons			863 *Lord Rodney* 302 /320 tons		
	Mins Secs		Speed	Mins Secs		Speed	Mins Secs		Speed
Sidmouth Jn	00.00		50½	00.00		45	00.00		
Honiton	07.25		36	08.45		25½	09.15		27½
MP 153 ½	09.15			11.40			11.50		
Seaton Jn	14.45		82	17.30		84½	17.20		86½
Axminster	17.15			20.00			19.50		
Hewish	29.15		47	33.30		40	33.20		36½
Crewkerne	31.10			35.35		75/55	35.35		77½/50½
Yeovil Jn	38.40		71½	43.30		67	43.50		71½
Sherborne	43.05		38/sigs	47.50		32/sigs	49.05		pws/21 pws
Templecombe	53.20		3E	57.40		1½ L	62.05		6L
	(52 net)			(56 net)			(56 ½ net)		

Down *Atlantic Coast Express*
749 *Iseult*
10 coaches, 321/335 tons

Location	Mins Secs	Speed		Mileage
Salisbury	00.00		T	0.0
Wilton	05.30			
Dinton	12.20			
Tisbury	17.10	55		
Semley	23.00	45		17.5
Gillingham	26.35	80		
MP 107½	28.40	pws 15		
Templecombe	34.35	62		28.4
Milborne Port	-	40		
Sherborne	40.55	83		
Yeovil Junction	44.30	55		39.1
Crewkerne	52.45	70		47.9
Hewish (MP 133¼)	55.15	38½		
Chard Junction	60.55			
Axminster	64.50	82		61.0
Seaton Junction	67.25			
Honiton Tunnel	74.25	28		
Sidmouth Junction	83.00 (80½ net)		3E	75.8

749 *Iseult*, a regular performer between Exeter and Salisbury in the 1920s, at Exeter Queen Street (later Central) with an up West of England – Waterloo express, 3 August 1928.
(H.C. Casserley)

738 *King Pellinore* on a 13-coach up West of England express descending Honiton bank at Seaton Junction, c1930.
(Colin Garratt Collection/Rev A.W. Mace)

not match the GWR 4-cylinder locomotives, which averaged 80,000 miles before an Intermediate overhaul, and in excess of 150,000 miles between 'Heavy Generals'. It is hard to make comparisons, however, without knowing the exact specification of the type of repair necessary at these two levels of Works attention.

The N15 boilers were relatively trouble free from a maintenance standpoint, although motion wear was extensive and the 2-cylinder 4-6-0s could become rough-riding after around 25,000 miles.

The allocation of the N15s at the start of the Second World War in September 1939 was:

Nine Elms: 736-742
Salisbury: 745-748
Exmouth Junction: 743, 744
Eastleigh: 749-755

In 1940-1, Bulleid fitted five N15s with Lemaître blast-pipes and large diameter chimneys. 755 *The Red Knight* was the only N15 to have retained 22in cylinders and, now fitted with double-ported piston valves and the new exhaust system, it built a considerable reputation and was transferred to Nine Elms (in exchange for 740), where it was said not only to be better than the other N15s (including the other four with the Lemaître exhaust) but was also better than that depot's King Arthurs and Lord Nelsons. The four Lemaître N15s performed better than the other standard N15s, but could not match 755. During the war period and up to nationalisation it shared top-link work with Nine Elms' Lord Nelsons until both were displaced by the new Bulleid Pacifics.

Many transfers took place during

Urie N15 744 *Maid of Astolat* climbing Honiton Bank with a West of England express, 1934. (ColourRail)

2-cylinder passenger engines were preferred to the more complex three and 4-cylinder locomotives, with Lord Nelsons in particular often being substituted by N15s or King Arthurs.

In October 1942 additional freight motive power was needed on the LNER, and the Southern was directed to send ten 4-6-0s on loan. They attempted to get rid of ten Lord Nelsons (852, 853, 857-859, 861-865) but as soon as their 4-cylinder motion was made known, they were rejected and the Ministry of War Transport Division ordered the SR management to send ten N15s instead. The engines chosen were 739, 740, 742, 744, 747-751 and 754. They were allocated to Newcastle (Heaton) and shared duties with LNER classes B16 4-6-0s and K3 2-6-0s, roving to Edinburgh, Leeds, Hull and Doncaster, mainly on freight work,

the war years, with Eastleigh boosting its allocation to twelve (739, 740, 742, 744, 747-754) to cover ten diagrams, mainly goods or van trains to and from Southampton Docks, with just a couple of main line semi-fast passenger services. Towards the end of the war, engine maintenance was increasingly difficult and the simple and robust

738 *King Pellinore* at Waterloo on an express for Bournemouth, 21 March 1936. (John Scott-Morgan Collection/H.F. Wheeler-R.S. Carpenter)

747 *Elaine* tender first with a two-coach local train, c1946. (MLS Collection)

but with occasional passenger train forays to Harrogate, Scarborough, Darlington and Doncaster. On 18 December 1942, 742 *Camelot* was called on to rescue the 10am Edinburgh–Kings Cross wartime *Flying Scotsman* when an A3 failed at Berwick, and took the twenty-coach train as far as Newcastle, with only nine minutes loss of time. 751 *Etarre* was also noted several times in March 1943 piloting a D21 4-4-0 on the evening Newcastle–Liverpool express. The N15s are reported to have been reasonably well received by the foreign crews – perhaps the Scots crews felt at home in the Drummond cabs! All were eventually replaced by the USA S160 2-8-0s, which were on loan to UK railways before being needed on the continent, and they returned to the Southern in July 1943, all

except 744, going to Eastleigh; 744 returned to Salisbury and resumed work over the Salisbury–Exeter section, its old 'stomping ground'. As D-Day approached, they found themselves, along with the H15s, increasingly moving the troops into position ready for embarkation for the Invasion.

As the sphere of war activity moved to France and Germany, passenger travel restrictions were relaxed. In the summer of 1945, the Southern could not cope adequately with the huge number of passengers seeking relief after the stresses of the war. The Urie engines returned to their depots of yore to undertake passenger duties once more. Salisbury received 744-746, Exmouth Junction 738 and 753, and Nine Elms hung on to 755. The rest remained at Eastleigh but were

increasingly on holiday specials to Bournemouth instead of their previous freight work.

The passenger livery of malachite green was restored in 1946, the first N15 to receive it appropriately its 'star', 755. By the end of 1947 most of the N15s were in the Bulleid livery although seven of the class did not go through Eastleigh Works for general overhaul until after nationalisation, so received the Southern livery with their new BR numbers. O.S. Nock was making a number of footplate trips in the immediate post-war period and around 1946 – Nock doesn't quote the date in his book *The Southern King Arthur Family* (published by David & Charles in 1976) – and was fortunate to experience a recently ex-works Urie N15, 747 *Elaine*, with a really heavy load of sixteen Maunsell coaches, weighing 443 tons tare/495 tons gross (the train was packed) on the 11.30am Waterloo–Bournemouth, for many years an Eastleigh diagram. Nock noted that it was fitted with a Maunsell 200lb boiler and made an excellent run with this load, albeit the driver certainly knew he was under scrutiny from an expert who would be publishing the log and making comment. After a slow start and a bad signal check at Wimbledon, 27% cut-off and almost full regulator got the train up to 66mph by Esher, and afterwards eased over the MP 31 milepost at 44mph. Nock comments that the coal was poor and with pressure falling, the easement after Woking was necessary, but time was well in hand. Pressure fell to 160lb, but with very restrained running after MP31, Basingstoke was reached

over a minute early. Departure from Basingstoke was vigorous and made without the trace of a slip, and the summit at Litchfield Tunnel was passed at 53mph. After the Micheldever stop, 747 was opened out and thundered down to the Winchester stop with full regulator and 27% cut-off once more, attaining a maximum speed of 75mph. Another 72 after Eastleigh and Southampton was reached 1½ minutes early. The terrific exertion between Micheldever and Winchester recorded an exceptional 1,500 indicated horsepower, though it is doubtful whether the locomotive could have sustained

this for long.

In 1947, Britain had a long icy winter coinciding with a severe coal shortage, so the Government instigated the conversion of a fleet of locomotives on the GWR and SR to oil-burning. Once more the Urie N15s were involved and 740, 745, 748, 749 and 752 were converted, but the change was short-lived as within a year the Treasury was complaining of a balance of payment problems causing a shortage of currency to buy the necessary oil. 740, the first engine to be converted in December 1946, gave some trouble initially when coasting with the regulator closed

and in August 1947 its blastpipe diameter was reduced by half an inch to increase the draught. It then performed successfully on the 7.20am Eastleigh–Waterloo and 11.30am return through to Bournemouth. The other oil-burning engines were then similarly modified. Whilst popular with the crews (especially the firemen!), they were not appreciated by the passengers in the first coaches because of the oily fumes. All the oil-burning engines were shedded at Eastleigh and worked Waterloo-Bournemouth services, plus semi-fast and parcels services on the Bournemouth and Waterloo–

A multiple jet exhaust Urie N15, believed to be 755 *The Red Knight*, at speed near the summit of Milepost 31, between Brookwood and Farnborough, with a Waterloo–West of England express, c1947. (ColourRail)

Salisbury routes. The last recorded sighting of an oil-burner before reconversion was of 745 on 2 October 1948 at the head of an up Bournemouth relief train.

The allocation of the N15s in the last year before nationalisation (at January 1947) was:

    Nine Elms: 738, 753, 755
    Salisbury: 744-748
    Eastleigh: 737, 739-742, 749-752, 754
    Bournemouth: 736, 743 (again!)

As experienced earlier by Nock, the Eastleigh engines regularly worked the 11.30am Waterloo–Bournemouth and the 5.50pm Bournemouth–Waterloo and the majority of the Southampton Dock boat trains. The Salisbury engines in the summers of 1947 and 1948 were noted working relief West of England holiday expresses right through from Exeter to Waterloo. Although time was seldom kept (with the usual summer Saturday delays), 746 was said to be a particularly reliable performer. Just after nationalisation, during the 1948 locomotive exchanges, 753 took over the packed 12-coach up *Atlantic Coast Express* after the failure at Salisbury of the A4 *Mallard*, and actually regained nearly six minutes of the time lost by the engine change.

9.3: Maunsell's King Arthurs

The Southern Railway Board began to accelerate its services in the late 1920s and Bournemouth two-hour trains were reintroduced in 1929, with much heavier loads than had been the custom before the First World War. The 5.15pm from Bournemouth was scheduled non-stop to London in exactly the two hours in the 1929 timetable and an early run that held schedule exactly is logged below.

This run started typically slowly with scrupulous observance of the line speed restrictions, dropping a couple of minutes to Southampton, then, after a good climb to Litchfield, was checked at Worting Junction and passed Basingstoke four minutes late. There followed a glorious sprint for home, unchecked, recovering all the lost time. It will be noted that all the passing times are to the nearest five seconds, typical of many train recorders in that era.

Other runs on the new Bournemouth two hour trains were highlighted in the 1929 *Railway Magazine*. 788 *Sir Urre of the Mount* with 375 tons on the 12.30pm Waterloo, got to the Southampton stop in 90 minutes 15 seconds (net 86¾) with nothing over 65mph before Woking, 49 minimum at

5.15pm Bournemouth – Waterloo
774 *Sir Gaheris*
11 coaches 356 / 370 tons

Location	Mins Secs	Speed	Schedule	Mileage
Bournemouth	00.00			0.0
Pokesdown	04.45	64½		1.8
Christchurch	06.45	45*		3.7
Hinton Admiral	11.05	38½		6.9
New Milton	14.45			9.4
Brockenhurst	21.15	70½		15.2
Beaulieu Road	25.45	60		19.9
Lyndhurst Road	28.20	66		22.6
Totton	31.05	30*		25.4
Southampton West	36.05	46	34½	28.7
Northam Junction	37.55	20*	36½	29.8
Eastleigh	44.50	49	43	34.4
Winchester	53.05	51½/50½		41.3
Micheldever	63.20	48		49.8
Litchfield	65.50	48		51.8
Worting Junction	73.15	sigs 30*	69½	
Basingstoke	75.55	66/70½	72	60.1
Hook	80.50	65		65.7
Fleet	86.00	71½		71.4
Milepost 31	90.45	70½		77.0
Woking	96.10	77½	93	83.6
Byfleet	98.15	80½		86.3
Weybridge	-	71½		
Walton	101.55	75		90.9
Hampton Court Junction	104.55	77½	103½	
Surbiton	106.00	71½		95.9
Wimbledon	110.05	68		100.7
Clapham Junction	113.25	35*	113	
Waterloo	<u>120.00</u>		<u>120</u>	<u>108.0</u>

Milepost 31, 61½ at Hook, 45½ at Worting and 75 at Winchester followed by a p-way slack at Swaythling. 792 *Sir Hervis de Revel* with 370 tons was faster, clearing Surbiton in 15¾ minutes after a vigorous start, ready for a p-way slack at Esher, then 61½ at Woking with only a drop to 56 at Milepost 31, 68 at Fleet and the low 70s down the bank to Winchester, which saw 792 into Southampton in 87 minutes 50 seconds (net 84½) against the 89-minute schedule.

The Southern Railway, after the 1929 tentative steps towards restoration of the Bournemouth two-hour trains, had inaugurated a number of return two-hour runs by

1930 and the King Arthurs of Nine Elms and Bournemouth reigned supreme on these until Schools were released from the Portsmouth services on electrification in 1937. The trains were scheduled for a maximum of eleven coaches, 365 tons. The *Railway Magazine* printed a couple of down runs and one up in 1931. (See below left) 452 made a very fast start to MP 31 and then had so much time in hand that it eased right back. Bournemouth's 789 had a packed Christmas Eve train made up to the maximum load and would have kept time without the bad signal checks around Southampton, possibly from relief trains running ahead.

The correspondent who sent these logs to Cecil J. Allen commented that the King Arthurs were preferred to the Lord Nelsons on the Bournemouth non-stops as there was concern on whether the water supply would hold out over the entire length

455 Sir Launcelot with a Waterloo–Salisbury train, June 1934.
(J.M. Bentley Collection)

Down Bournemouth 2-hour Non-Stop Expresses					
452 *Sir Meliagrance*			789 *Sir Guy*		
10 coaches, 317/335 tons			13 coaches, 422/460 tons		
Location	Mins Secs	Speed		Mins Secs	Speed
Waterloo	00.00			00.00	
Vauxhall	03.30			04.00	
Clapham Junction	06.45			07.45	
Wimbledon	10.20			12.10	
Surbiton	14.55	70		17.30	65½
Weybridge	21.18	68/75		24.15	70
Woking	26.00			29.20	65½
MP 31	32.32	60		36.45	53
Fleet	34.37	69		38.50	70
Winchfield	41.05	eased		45.05	67½
Basingstoke	50.05	54		52.25	
Worting Junction	52.50			55.20	47/39
Micheldever	62.00			65.20	
Winchester	69.50	70		72.20	77½
Eastleigh	76.35	pws		78.10	sigs
Southampton	83.40			90.40	sigs
Totton	88.15			95.30	
Beaulieu Road	95.12			102.40	
Brockenhurst	99.45			107.25	64½/49
Hinton Admiral	-			-	70
Christchurch	111.25			119.10	
Bournemouth Ctl	116.45	3E		124.00	4L
	(114 net)			(118 net)	

Up Bournemouth 2-Hour Non-Stop Express
775 *Sir Agravaine*
11 coaches, 357 / 375 tons

Location	Mins Secs	Speed
Bournemouth Ctl	00.00	
Christchurch	06.21	64½ / 45*
Hinton Admiral	-	53
Brockenhurst	18.26	66
Beaulieu Road	22.50	60
Lyndhurst Road	-	69
Totton	27.43	sigs 15*
Southampton	32.03	sigs
Eastleigh	43.00	42
Winchester	52.02	48½
Micheldever	62.11	50½
Worting Junction	71.43	sigs 30*
Basingstoke	74.34	
Winchfield	81.55	65
Fleet	87.51	
MP 31	90.00	
Woking	95.38	
Weybridge	99.52	75
Surbiton	105.53	70
Wimbledon	110.10	sigs
Clapham Junction	114.03	
Vauxhall	118.03	
Waterloo	121.18	(116 net) 1¼ L

772 **Sir Percivale** at Bournemouth Central with the up *Bournemouth Belle* Pullman train, while Urie N15 749 *Iseult* stands in the bay, ready to take over a following Weymouth–Waterloo express, 1936. (ColourRail)

without replenishment on the larger engines. The King Arthurs clearly had time well in hand on these services unless the load was significantly increased.

Although the King Arthurs rode well when newly out of shops, they could get a bit rough as their mileage rose. They had some fast freight workings also and O.S. Nock had a footplate pass for the overnight Exmouth Junction–Nine Elms fast freight in the late 1920s. From Exeter he had new Maunsell S15 826, with the light load, made up to ninety axles from Templecombe, and the run to Salisbury is described in Chapter 9.5 later. At Salisbury, Nine Elms' 778 took over, Driver Gray of Nine Elms warning Nock that he was in for a rough ride, and Nock's log is shown below.

After the climb up Porton Bank, Nock states that the cut-off was fixed at 25% and Driver Gray drove on

Salisbury – Nine Elms Express Freight
778 *Sir Pelleas*
90 axles, 500 tons gross (= 600 tons of passenger bogie stock)
Driver Gray, Fireman Benton, Nine Elms

Location	Mins Secs	Speed
Salisbury	00.00	
Porton	14.52	25
Grateley	27.33	35
Red Post Junction	-	60
Andover Junction	39.15	pws
MP 62½	-	25
Overton	-	41
Oakley	65.54	48
Basingstoke	71.55	58
Farnborough	90.53	41*
Woking	102.28	52
Surbiton	119.28	sigs
Wimbledon	129.40	
Clapham Junction	137.12	sigs
Nine Elms	147.20	(130 net) Sch 134 (38 mph average)

Up *Southern Belle*
803 *Sir Harry le Fise Lake*
12 coaches (incl Pullmans) 383 / 410 tons

Location	Mins Secs	Speed	Schedule	Mileage
Brighton	00.00			0.0
Preston Park	03.40			1.3
Clayton South End	09.05	42		4.7
Hassocks	11.45			7.1
Keymer Junction	14.00	72	13	9.8
Haywards Heath	16.50		16	12.9
Balcombe	20.45			15.8
Balcombe Tunnel	23.05	54		19.0
Three Bridges	25.20		25	21.4
Horley	28.25	70		24.9
Earlswood	32.30	53	32½	29.0
Quarry Box	36.05	47		32.1
Star Lane Box	38.30			34.1
Purley	41.50	64½		37.3
East Croydon	44.50		46	40.4
Streatham Common	49.20			44.3
Balham	52.00		52	46.0
Clapham Junction	54.55	sigs 20*	55	48.2
Wandsworth	-	sigs 20*		
Victoria	61.20		60	51.0

announcement during the event of the approval of electrification to Brighton. 795 *Sir Dinadan* and 804 *Sir Cador of Cornwall* hauled the last down and up *Southern Belle* Pullmans respectively on 31 December 1932, the day before the new electric services to Brighton were inaugurated. A representative King Arthur run on the *Southern Belle* was included in Cecil J. Allen's *Locomotive Practice and Performance* article in the March 1929 *Railway Magazine*. (See left)

Typically the start from Brighton to Clayton was slow compared with the Baltic tanks but livened up after Keymer Junction, with the climb from Horley to Quarry Box being excellent. The net time was just a minute under schedule. Coal consumption of 803 during a month in 1929 averaged 38lb per mile compared with the Marsh Atlantics' 40lb.

By the early 1930s the Central Section drivers had got the measure

767 *Sir Valence* starts away from Brighton with the 5.5pm *Southern Belle* Pullman train, 30 April 1932. A 'Gladstone' 0-4-2 appears to be standing on the right.
(J.M. Bentley/H.C. Casserley)

the regulator. The engine steamed freely, and rode comparatively well, but the racket set up by the loose cab fittings was deafening – described as 'earsplitting'.

Pullman services were reintroduced to the Southern Railway's Western Section in July 1931 with the inaugural *Bournemouth Belle* being hauled by 780 *Sir Persant*. The Nine Elms engines, 773-782, normally worked this service on weekdays with Lord Nelsons taking charge of the heavier weekend loads. The six-wheel tender Arthurs worked Pullman trains on the Central Section too and 793 *Sir Ontzlake* hauled the celebratory twenty-fifth Anniversary *Southern Belle* on 1 November 1929, with

Victoria to Brighton

	802 *Sir Durnore* 333 / 360 tons		796 *Sir Dodinas le Savage* 424 /455 tons		805 *Sir Constantine* 404 / 440 tons	
Location	Mins Secs	Speed	Mins Secs	Speed	Mins Secs	Speed
Victoria	00.00		00.00		00.00	
Clapham Jcn	05.40		06.05		05.21	46
Balham	08.20		09.00		08.03	42
Streatham C'mon	10.40		11.50		10.35	53½
Selhurst	14.00		15.05	52	13.48	49
East Croydon	15.20	sigs	16.35		15.08	
Purley	20.40	sigs	20.35	sigs	18.45	53
Quarry Box	29.15		30.10	pws	25.07	48
Earlswood	32.15		34.55		27.58	
Horley	35.45	74	39.00	75	31.33	74
Three Bridges	38.55		42.15		34.59	
Balcombe Tunnel	41.15	57	44.50	50	38.11	44
Balcombe	43.20		47.05		40.41	
Haywards Heath	46.35	76	50.20	78	44.26	69
Hassocks	51.25	54	55.25	52	49.47	53½
MP 46	53.45		57.50		52.25	
Preston Park	57.30	61	61.15		55.57	sigs
Brighton	60.20		64.20		59.14	
	(56¼ net)		(58¾ net)		(58½ net)	

of the King Arthurs and good runs were being consistently achieved with some fairly substantial loads on the standard 60-minute non-stop schedule to Brighton. Above are three such runs logged in the pages of the 1931 *Railway Magazine*.

In the up direction the article described seven runs with loads varying from 360 to 430 tons all with net times between 56½ and 59 minutes. Four journeys were on time or slightly early, two were one to two minutes late and one was five minutes late.

There was a final flurry of interest just before the electrification, and one very fast up run was tabulated (shown right).

The same article reported a run with 763 *Sir Bors de Ganis* with a heavier load of 385 tons which

Brighton–Victoria, 1932
766 *Sir Geraint*
10 coaches, 310 tons

Location	Mins Secs	Speed
Brighton	00.00	
Preston Park	03.29	
Clayton Tunnel	-	50
Hassocks	10.22	
Keymer Junction	12.35	76
Haywards Heath	15.09	
Balcombe	18.53	
Balcombe Tunnel Box	21.04	56
Three Bridges	23.17	
Horley	26.02	79
Earlswood	29.40	
Quarry Box	32.46	51
Purley	38.19	68
East Croydon	41.42	sigs
Balham	48.37	sigs
Clapham Junction	51.43	sigs
Victoria	58.13	1¾ E (52½ net)

Down *Atlantic Coast Express*

Location	784 *Sir Nerovens* 279 / 285 tons		456 *Sir Galahad* 352 / 370 tons		784 *Sir Nerovens* 407 / 430 tons		850 *Lord Nelson* 389 / 410 tons	
	Mins Secs	Speed	Mins Secs	Speed	Mins Secs	Speed	Mins Secs	Speed
Salisbury	00.00		00.00		00.00		00.00	
Wilton	06.30		05.55		06.45		06.25	
Dinton	12.40	65	12.20	57 ½	14.00	54	15.55	pws
Tisbury	16.40	62	17.05	48	19.10	54	21.25	50
Semley	21.30	54	25.00	pws	25.20	42	27.35	46
Gillingham	25.15	77½	29.19	72	29.10	70	31.00	80
T'combe	33.10	pws/61	35.45	53½/71½	37.20	pws60	37.10	54/72
Sherborne	40.05	35/82	42.20	40½/75	44.20	34½/79	43.15	48/79
Yeovil Jcn	43.42	47	46.20	45	48.05		46.45	50
Crewkerne	52.40	70½	55.25	67	56.25	71½	55.20	75
Hewish	54.50	40½	58.10	36½	59.15	33	58.05	32
Axminster	64.40	80	67.45	77½	69.30	83½	68.00	85
Seaton Jcn	67.15		70.25		72.10		70.30	
MP 153½	78.15	21½	80.30	26¾	83.15	20½/24½	81.35	21
Honiton	79.50		81.55	77½	84.50	pws	83.10	70
Sidmouth Jn	85.20	2E	87.05	T	90.30	3L	86.55	pass
	(83 net)		(84½ net)		(88 net)		(84½ net to stop)	

completed the run in 58 minutes 14 seconds, net 56½ minutes.

The King Arthurs rode reasonably well for a 2-cylinder engine, especially those attached to the bogie 5,000 gallon tenders, though not as smooth riding as the 4-cylinder Lord Nelsons. They could get rough as the mileage mounted, but drivers were not afraid to work them up to very high speeds, especially on the switchback route west of Salisbury and high 80s were common with the occasional 90mph. As indicated in the previous chapter, 91mph with *Sir Lamorak* is the highest known speed recorded with a King Arthur in the pre-war period.

Articles in the 1930 editions of the *Railway Magazine* picked up the accelerated timing of the Atlantic Coast Express, booked from Salisbury to Sidmouth Junction non-stop in 87 minutes in the Down

direction. This was at a time that King Arthurs and Lord Nelsons were used indiscriminately on Nine Elms top link duties east of Salisbury and I quote four runs west of Salisbury below, three with

King Arthurs and one with a Lord Nelson, which was unusual on that section of line (see above)

The *Railway Magazine* did not show speeds for 850's run, so I have estimated speeds in comparison

Six-wheel tender Arthur, 793 *Sir Ontzlake*, on the up *Southern Belle*, c1932.
(J.M. Bentley/O.J. Morris)

Down *Atlantic Coast Express*
10.40am Waterloo – Exeter Relief

Location	778 *Sir Pelleas* 279 / 295 tons			862 *Lord Collingwood* 284 / 300 tons			773 *Sir Lavaine* 389 / 410 tons		
	Mins Secs		**Speed**	**Mins Secs**		**Speed**	**Mins Secs**		**Speed**
Waterloo	00.00			00.00			00.00		
Vauxhall	03.43			03.40					
Clapham Jcn	07.19			07.00			07.13		
Wimbledon	11.18		sigs	10.50			10.53		
Surbiton	16.49		sigs 25	16.00		pws Esher	15.48		68
Weybridge	25.33		66	24.15			22.32		75
Woking	30.29			29.15		66	27.32		sigs (boat trains)
MP 31	37.08		58	36.05		55	39.00		sigs
Farnborough	39.17			38.20		64	41.38		
Fleet	42.16			41.25			44.53		
Hook	47.20		70	49.00		pws/60	50.29		65
Basingstoke	52.12			54.45			56.24		sigs
Worting Jcn	54.38		58	57.50		50	61.55		
Overton	59.41			63.30			68.36		
Hurstbourne	64.11		75	68.30			73.22		72
Andover Jcn	68.28		77½	73.00		75½	77.32		80
Grateley	74.39		43	79.10		52	83.22		48
Porton	80.53		70	84.50		80	89.06		76½
Tunnel Jcn	85.18			88.20			92.50		
Salisbury	87.52		2E	90.10		T	94.50		2¾ L
	(85½ net)			(87½ net)			(88 net)		

with the other timings.

The *Railway Magazine,* in 1930, quoted some more King Arthur runs to Salisbury, a couple on the down ACE, and its heavier relief, and one with a Lord Nelson for comparison (see above).

Cecil J. Allen, in summarising these runs, wrote that every Lord Nelson feat is duplicated, if not improved on, by the best King Arthur runs. 778 above was eased considerably after Andover Junction, to avoid too early an arrival and consequent delays approaching Salisbury. In contrast, after a somewhat lethargic run, 862 hurried over the last stages in a desperate attempt to arrive on time and just about made it. 773, with a

train more than 100 tons heavier, virtually duplicated the Lord Nelson performance and made by far the best start of any of the runs, until it caught up a succession of boat trains and followed them until they cleared Worting Junction. It matches 862 in the latter stages of the run. The *Railway Magazine* logs are scanty in their speed indications and 773's log only gave the speeds at Grateley and Porton, so I've estimated some earlier speeds for the timings and my own experience.

During the early 1930s, the Nine Elms engine, 778 *Sir Pelleas,* was a reliable and excellent performer, recording a number of consistent runs on the *Atlantic Coast Express* in both directions. In December

1933, it was timed on the down run in 87 minutes 15 seconds with a load of 345 tons gross, beating the schedule by nearly three minutes, with a maximum of 82mph through Andover. On the same day in the up direction with 380 tons gross it beat the 92 minutes schedule by a full five minutes with maximum of 75mph at Andover and Woking to West Weybridge.

However, this was thoroughly eclipsed by the record-breaking run on the up *Atlantic Coast Express* with sister engine 777 *Sir Lamiel,* now thankfully preserved. Leaving Salisbury late with its winter load of 328 tons/345 tons gross, and with a remarkably clear road, it completed the 83.8 miles in 72 minutes 41

Up *Atlantic Coast Express*
777 *Sir Lamiel*
10 coaches, 328/345 ton

Location	Mins Secs	Speed	Mileage
Salisbury	00.00		0.0
Tunnel Junction	03.04	-	1.1
Porton	08.22	53	5.5
Grateley	14.00	66	11.0
Andover Junction	18.43	88½	17.4
MP 62 ½	-	70½	21.2
Hurstbourne	22.52	76½	22.7
Whitchurch	24.24	72½	24.6
Overton	27.20	72½	28.2
Oakley	29.54	79	31.4
Worting Junction	31.33	75	33.5
Basingstoke	33.26	88½	36.0
Hook	37.25	80	41.6
Winchfield	39.15	82	44.1
Fleet	41.33	85	47.3
Farnborough	44.00	80½	50.6
Milepost 31	-	76½	52.8
Brookwood	47.56	83½	55.8
Woking	50.26	88½	59.4
Byfleet	52.15	90	62.1
Weybridge	54.05	82	64.7
Surbiton	59.28	76½	71.8
Wimbledon	63.20	69	76.5
Clapham Junction	66.27	40*	79.9
Waterloo	72.41		83.8

777 *Sir Lamiel* passing Vauxhall with the 3pm Waterloo – West of England, c1930.
(J.M. Bentley/Real Photographs)

777 *Sir Lamiel* heads a down West of England express past Raynes Park, c1932.
(MLS Collection)

seconds at an average speed of 71.6mph start to stop. Doubts have been expressed on the credibility of this exploit, but it was timed by a reliable recorder and if there was a following wind (a common phenomenon from the South West) the power output calculated – a sustained 1,500 indicated horsepower - was just within the range of the King Arthur class at its very best.

This famous log is recorded above.

About this time there were other splendid runs on the ACE, not least behind 776 *Sir Galagars* with

777 *Sir Lamiel* acts as standby for a royal train in the Up Bay at Woking, c1938. (John Scott-Morgan Collection/K. Nunn-LCGB Collection)

787 *Sir Menadeuke* works a Birkenhead–Bournemouth express of mixed SR/GW stock between Reading West and Basingstoke, c1935. (John Scott-Morgan Collection/M.W. Earley)

784 *Sir Nerovens* stands ready for departure with a Bournemouth–Waterloo express while Lord Nelson 864 *Sir Martin Frobisher*, newly modified by Bulleid, is ready to take over a Weymouth–Waterloo train, at Bournemouth Central c1939.
(Colin Garratt Collection/Rev A.W. Mace)

Up *Atlantic Coast Express*
776 *Sir Galagars*
13 coaches, 419/445 tons

Location	Mins Secs	Speed	Mileage
Salisbury	00.00		0.0
Tunnel Junction	03.49		1.1
Porton	09.50	39	5.5
Grateley	17.00	53½	11.0
Andover Junction	22.16	78½	17.4
Milepost 62 ½	25.34	62	21.2
Hurstbourne	26.51	68½	22.7
Whitchurch	28.35	64	24.6
Overton	31.53	65	28.2
Oakley	34.50	69	31.4
Worting Junction	36.39	71	33.5
Basingstoke	38.36	82½	36.0
Hook	42.46	78	41.6
Winchfield	44.38	76	44.1
Fleet	47.12	79	47.3
Farnborough	49.50	75	50.6
Milepost 31	51.39	72	52.8
Brookwood	53.59	80	55.8
Woking	56.41	80½	59.4
Byfleet	58.41	82	62.1
Weybridge	60.41	69	64.7
Esher	-	71	69.4
Surbiton	66.47	67	71.8
Wimbledon	71.21	58	76.5
Clapham Junction	75.07	40*	79.9
Waterloo	80.40		83.8

Exeter–Salisbury Up *Atlantic Coast Express*
450 *Sir Kay*
10 coaches, 331/355 tons

Location	Mins Secs	Speed
Exeter Central	00.00	T
Broad Clyst	08.15	69/40½
Sidmouth Junction	16.45	63
Honiton Tunnel West	24.17	34
Seaton Junction	29.45	82
Axminster	32.15	79
Chard Junction	37.12	
Hewish (MP 133¼)	44.22	45
Crewkerne	46.20	
Yeovil Junction	54.00	57
Sherborne	57.52	
Milborne Port	-	44
Templecombe	64.30	80
Mp 107½	68.10	60
Gillingham	70.12	71
Semley	74.40	42
Tisbury	79.47	
Dinton	83.30	72
Wilton	88.35	50*
Salisbury	92.20	T

a full summer load of 419/445 tons. It was obviously slower out of Salisbury up Porton Bank with this load, but sustained running at 78-82mph between Basingstoke and West Weybridge was first class. The full log is shown above.

The Exmouth Junction and Salisbury King Arthurs continued to give excellent service west of Salisbury and a typical run of that period and route was logged in a *Railway Magazine* article in 1933, shown above right.

After the Brighton line electrification in January 1933, the allocation of the Maunsell King Arthurs was as follows:

  Nine Elms: 452, 773-782
  Salisbury: 450, 451, 453-457
  Exmouth Junction: 448, 449, 769
  Bournemouth: 783-792
  Stewarts Lane: 763-768, 770-772, 793-796
  Ramsgate: 797-806

793-796 were usually rostered to the Victoria–Eastbourne services, with the Scotchmen concentrating on the Margate and Dover continental boat train routes.

In 1934, 770 and 771 were reallocated to Dover and were replaced at Stewarts Lane by the Nine Elms 'stars' 777 and 778. No further significant permanent transfers then took place until the Portsmouth route electrification in July 1937, when Schools 924-933 were sent to Bournemouth and replaced 783-792 on the main London services. Bournemouth retained 784 and 785 for the Poole–Basingstoke–Reading–Oxford–York and Birkenhead cross-country trains, but Exmouth Junction received 786-792, and 783 went to the South Eastern Section at Stewarts Lane. This then prompted a further stirring of the rest of the King Arthur fleet, viz:
  Nine Elms: 772-780 (777 and 778 returning 'home')
  Salisbury: 448-457

792 *Sir Hervis de Revel*, transferred to Exmouth Junction, is here seen on an up West of England express awaiting departure from Exeter Central, c1937. 792 was said to be the 'black sheep' of the Exmouth Junction King Arthurs and was fitted with a multiple jet exhaust and wide chimney by Bulleid, but this failed to improve its performance. It was only when that was replaced by a standard King Arthur chimney in the 1950s that 792's performance became equal to that of its peers.
(John Scott-Morgan Collection/J.M. Bentley)

Stewarts Lane: 763-769, 781-783, 793, 794, 798, 799
Ramsgate: 795-797, 800-806
Dover: 770, 771

The Stewarts Lane six-wheeled tender King Arthurs could also relieve the Marsh H2 Atlantics on the Newhaven boat trains if the loads were too heavy for the 4-4-2s.

The mileage run annually by the King Arthurs in the latter half of the 1930s was high, especially on the former L&SWR routes. Interestingly, the two Bournemouth engines, 784/5, achieved the highest of any, averaging 57,300 miles per annum, obviously working reliably day after day between Bournemouth and Oxford. Salisbury's Eastleigh Arthurs were next at 51,600, followed by Nine Elms' nine at just under 50,000 and Exmouth Junction's Scotchmen at 44,000. Because of the shorter routes, those on Eastern Section were significantly less at around 35-36,000. Coal consumption comparisons on passenger services were made in 1936 between the different Southern classes, and the King Arthurs were the most economical at 42½lb per mile, followed by the Lord Nelsons at 44lb; the N15s at 45¾lb; the Maunsell S15s at 46¾lb; the H15s at 47¼lb; the Urie S15s at 48lb and the N15X trailing at 51½lb.

Nine Elms had thirteen weekday King Arthur and N15 diagrams, but this surged to an impossible 31 summer Saturday turns. They would have to use H15s and Drummond T14s to fulfil their commitments. Salisbury had fourteen weekday and Saturday diagrams for the combined classes. Eastleigh had four weekday and six Saturday diagrams for Urie N15s, but these could be replaced by ex-works King Arthurs running in. On weekdays, two thirds of the Nine Elms turns were to Salisbury and one third to Bournemouth. Seven of the summer Saturday diagrams were scheduled right through to Exeter. The workings west of Salisbury were dominated by the Salisbury and Exmouth Junction King Arthurs, with support from the N15s and the Maunsell S15s. However, despite this official allocation of diagrams between the Urie N15s, the King Arthurs and Lord Nelsons, Western Section depots tended to use them turn and turn about, possibly selecting the most recently ex-works engine in preference or the locomotives with the best current reputation amongst the depot foremen and drivers. There were not enough Lord Nelsons for them to dominate any

Victoria – Margate Granville Express 797 *Sir Blamor de Ganis* 11 coaches, 361/385 tons			
Location	Mins Secs	Speed	
Victoria	00.00		T
Herne Hill	09.09	28*	
Sydenham Hill	12.57	26½	
Kent House	16.14	53	
Beckenham Junction	17.15	41*	
Bromley South	20.05	53	
Bickley Junction	22.55	34	2L
St Mary Cray	25.46	61½	
Swanley Junction	28.55	49	2L
Farningham Road	31.35	75	
Sole Street	38.55	46½	
Cuxton Road Box	-	68	
Rochester	47.09	pws 25*	
Chatham	48.11	¾ L	
Gillingham	50.52	35½	
Rainham	-	68	
Sittingbourne	61.07	72½	
Faversham	68.05	25	1L
Whitstable	76.10	64	
Herne Bay	80.13	46½	
Birchington	88.07	71½/48	
Margate	92.07 (90 net)		2L

Margate – Victoria 766 *Sir Geraint* 9 coaches, 290/310 tons			
Location	Mins Secs	Speed	
Margate	00.00		
Birchington	06.16	64	
Herne Bay	14.59	43½	
Whitstable	18.46	60	
Faversham	26.16	45*	1¼ L
Teynham	-	pws 30*	
Sittingbourne	35.44	55	3¾ L
Gillingham	45.38	40/64	
Chatham	47.54		4L
Rochester	48.54		
Cuxton Road Box	-	43	
Sole Street	59.54	40	3L
Farningham Road	66.19	76½	
Swanley Junction	69.07	53	2L
St Mary Cray	71.52	70½	
Bickley Junction	74.06	53½	1L
Bromley South	75.52	64½	
Beckenham Junction	78.45	40*	¾ L
Kent House	79.58	40*	
Sydenham Hill	83.04	46	
Herne Hill	85.18	40*	¼ L
Victoria	92.46	(90 net)	¾ L

route, so loads were limited to that which a King Arthur could manage. In any case, until Bulleid tackled the weaknesses of the 4-cylinder engines, many crews preferred a good King Arthur.

The King Arthurs dominated the Kent Coast and continental services with substitution of the 3-cylinder U1 Moguls on the Ramsgate trains and Schools on the continental services via Tonbridge when necessary. A couple of typical Kent Coast via Chatham runs were recorded in 1933 *Railway Magazine* articles (above and right).

The last five years of the 1930s before the start of the Second World War saw the heyday of the King

771 *Sir Sagramore*, with a six-wheel tender, passes class H 0-4-4T 329 at Paddock Wood with a continental express for Dover Marine, 17 October 1929. (J.M. Bentley Collection/H.C. Casserley)

770 *Sir Prianius* with an eight-wheel flush-sided tender coming off the Folkestone Harbour branch with a boat train for Victoria, c1930. (J.M. Bentley Collection)

Arthurs, especially on the West of England route, while the Schools had the best trains to and from Bournemouth. 768 *Sir Balin*, that earned fame in the 1950s when it was Sam Gingell's regular engine at Stewart's Lane, was transferred from the Kent scene to Exmouth Junction in the early 1930s and was fitted for a time with a large 5,000 gallon flat-sided tender similar to that of the Lord Nelsons. It was credited with many fine runs between Salisbury and Exeter, of which the one tabled below, with Driver Young, is an excellent example. I include also an excellent run with another grand performer, a Nine Elms engine, 779, leaving Salisbury seventeen minutes late and recovering ten of them with a load of twelve coaches (see above).

453 *King Arthur* himself with 450 tons gross gained six minutes on the 98-minute schedule with 45 at Semley, and 27 at Honiton, with 80s

795 *Sir Dinadan* on a continental boat train c1929. (Colin Garratt Collection/Rev A.W. Mace)

763 *Sir Bors de Ganis* near Dumpton Park with a Ramsgate express, c1935. (Colin Garratt Collection/Rev A.W.Mace)

Down *Atlantic Coast Express*

	768 *Sir Balin* 13 coaches, 421/455 tons			779 *Sir Colgrevance* 12 coaches, 388/415 tons		
Location	**Mins Secs**	**Speed**	**Mins Secs**	**Speed**		**Mileage**
Salisbury	00.00		00.00		17L	0.0
Wilton	06.15		05.35			2.5
Dinton	13.05	59	11.40			8.2
Tisbury	17.50	53	15.55	63		12.4
Semley	23.50	42	21.05	52		17.5
Gillingham	27.30	82	24.25	82		21.6
Milepost 107 ½	29.19	64	26.18	63		23.9
Templecombe	32.55	82	29.55	82		28.4
Milborne Port	35.30	50	-	51		30.8
Sherborne	38.45	85	35.37	85		34.5
Yeovil Junction	42.10	77	39.10			39.1
Sutton Bingham	44.05	54	-	56		41.3
Crewkerne	50.15	71	47.05	71		47.9
Milepost 133¼	52.45	37	49.15	42		49.7
Chard Junction	58.25	80	54.50			55.9
Axminster	62.00	86½	58.40	79		61.0
Seaton Junction	64.30	71	61.20			64.2
Milepost 152½	71.30	26½	-	25		69.0
Milepost 153½	73.31	32	-	29		70.0
Honiton	74.50	62	72.25			71.2
Sidmouth Junction	78.35	82/65	76.10	80		75.8
Whimple	81.50	76	-			79.5
Broad Clyst	84.30	83	82.00	80		83.2
Exmouth Junction	87.45		-	sigs		
Exeter	90.00		88.10 (87¾ net)		7L	88.0

769 *Sir Balan* with a 5,000 gallon bogie tender on a Dover – Victoria train in Folkestone Warren, 26 April 1938. (J.M. Bentley Collection)

at Sherborne, Axminster, Sidmouth Junction and Broad Clyst (86).

The running on the Bournemouth line got better too, possibly the King Arthurs competing with the Schools that would mainly replace them on the two-hour trains. The 1934 *Railway Magazine* recorded one exceptional run with a King Arthur on a 12-coach train that completed the run well within schedule and 113¼ minutes net.

An Up run on a Bournemouth two-hour train was briefly noted in 1937 with 787 *Sir Menadeuke* on

771 *Sir Sagramore*, now once more with a bogie tender, passing through Folkestone Warren with a Dover – Victoria express, 26 April 1938. (J.M. Bentley Collection)

781 *Sir Aglovale* on a Victoria–Dover boat train, c1935.
(Colin Garratt Collection/Rev A.W. Mace)

10 coaches, 323/345 tons which reached Waterloo in 111 minutes net, with an excellent climb to Litchfield summit , accelerating from 44mph at Winchester, to 49 at Winchester Junction; 53 at Wallers Ash; 57 at Micheldever; 58 at Litchfield Tunnel; 67½ at Wootton Box and 80 through Basingstoke.

With electrification as its priority and war looming, Bulleid, on taking over from Maunsell in 1937, found that investment in new steam locomotives was very limited and King Arthurs could perform reliably with just routine maintenance. Bulleid saw them as simple and robust but rather outdated machines and, apart from tinkering with multiple jet exhausts for a few engines, he saw sorting the Lord Nelsons out as his priority with the limited cash available before he persuaded his Board to invest in the revolutionary designs he was contemplating.

At the beginning of the Second World War, the ten Lord Nelsons on the Eastern Section found themselves unwanted as all continental traffic virtually came to a halt, and the need was for troop trains and freight, for which the King Arthurs were much more suitable. Therefore 772-774, 777-780 were transferred from Nine Elms to Stewarts Lane, 784 and 785 went from Bournemouth and 795 from its activities on the former Central Section. With this influx, there was a redistribution of the King Arthurs already on the Eastern Section, with 764 and 767 to Dover, 798 and 799 to Ramsgate and, because of the extra freight movements, 765, 768, 769, 794, 796, 801 and 806 to Hither Green. These replaced their

**Down *Bournemouth Limited*
784 *Sir Nerovens*
12 coaches, 385/415 tons**

Location	Mins Secs	Speed	
Waterloo	00.00		T
Queens Road	-	sigs	
Clapham Junction	07.30		
Wimbledon	12.20	sigs	
Surbiton	19.45	pws	
Walton	24.30	67½	
Byfleet	28.43	70½	
Woking	31.00		
MP 31	37.30	58	
Fleet	42.40	73½	
Basingstoke	52.15	66	
Worting Junction	54.47		
Micheldever	63.13		
Winchester	70.20	77½	
Eastleigh	75.52		
Southampton	82.58		
Lyndhurst Road	92.28		
Beaulieu Road	95.50		
Brockenhurst	100.38		
Hinton Admiral	109.20	75½	
Christchurch	112.05		
Bournemouth Central	116.32 (113¼ net)	3½ E	

five S15s which were required
at Feltham for heavy freights to
Southampton Docks.

After the initial flux of troop
movements, and more changes, the
wartime allocation of King Arthurs
settled down as follows:

Nine Elms: 775-778, 781 and 782
Salisbury: 448-457
Exmouth Junction: 786-792
Bournemouth: 774
Stewarts Lane: 763-773, 780, 785,
795, 799
Hither Green: 779, 783, 784, 794,
796-798
Ramsgate: 800, 802-805

On the Western Section, Nine
Elms, Salisbury and Exmouth
Junction retained their most
recently overhauled King Arthurs
for the remaining passenger turns
and those examples longer out of

shops, and with higher mileage
since repair, took over some of
the freight work of the H15s and
S15s, releasing those for additional
military goods traffic to the ports.
Mileages between heavy repairs
were lengthened to 125,000,
making more available for freight
traffic, albeit in a more run down
condition.

A few trains from the channel
ports to the north avoiding London
were initiated after the immediate
threat of invasion had become
less likely, using a Ramsgate King
Arthur to Ashford and a U1 Mogul
to Reading before changing to GWR
motive power to Banbury and then
LNER engines.

As the war progressed, more
passenger services were withdrawn
and those that remained became
heavier and slower, with more

stopping places. This caused
problems for the Schools on
the Bournemouth run and 773,
779 and 781 were drafted to
Bournemouth depot in July 1941
to help out. Later, Nelsons 850-855
were released by Nine Elms for
Bournemouth. Then the Southern
was required to send ten of the Urie
N15s to the north-eastern section of
the LNER rather than the surplus
Nelsons, so 772, 778, 793 and 795
were dispatched to Eastleigh to
replace some of its stud.

There are very few logs of
King Arthur runs during the war
for fairly obvious reasons, but
there was one in a 1941 *Railway
Magazine* with one of the Salisbury
Eastleigh Arthurs on an Exeter–
Waterloo service that stopped
only at Andover and Woking east
of Salisbury. It cannot perhaps be

450 *Sir Kay* removes
a parcels vehicle from
an ECS/van train at
Eastleigh, c1939.
(MLS Collection)

Salisbury – Woking, 1940 or 1941
448 *Sir Tristram*
11 coaches, 355 / 380 tons

Location	Mins Secs	Speed
Salisbury	00.00	
Tunnel Junction	03.46	
Porton	10.21	42/38
Grateley	17.45	52
Red Post Junction	-	80
Andover Junction	23.20	
	00.00	
MP 62 ½	-	38
Hurstbourne	09.14	53
Whitchurch	11.25	51
Overton	15.12	62
Oakley	18.20	65
Worting Junction	20.20	sigs 20*
Basingstoke	23.14	
Hook	30.40	
Winchfield	33.07	71
Fleet	36.07	
Farnborough	38.58	
MP 31	40.56	64
Brookwood	43.33	75/sigs
Woking	47.47	(43½ net)

801 *Sir Meliot de Logres* hauls a post-war freight through Folkestone Warren, 1947. (J.M. Bentley Collection)

compared with their best work before the war, but it is of interest that fairly normal speeds were maintained (see above).

During the middle period of the war, the Southern management eased some of the locomotive route restrictions, as happened on other railways, to increase the flexibility to cover essential freight working. King Arthurs with 5,000 gallon tenders were permitted to work freight on the Brighton and other Central Section routes, although the authorities had to be careful to only roster six-wheeled tender engines to diagrams involving the use of the short Three Bridges turntable.

Then the return of the loaned Urie N15s caused another upheaval in King Arthur allocation, which in

July 1943 was as follows:

Nine Elms: 766-768, 770, 771, 786, 780 and 788
Salisbury: 448-457 as before with the addition of 773 and 774
Exmouth Junction: 775, 786, 787 and 789-792
Bournemouth: 772, 777 and old favourites, 784 and 785
Stewarts Lane: 763-765, 769, 778, 779, 781-783, 793-796
Ashford: 797-801
Hither Green: 802-806

The Ashford-allocated engines replaced the H2 Marsh Atlantics on slow passengers and freights for which they were totally unsuited and the 3-cylinder N1s displaced 802-806, which moved to Ashford also, their place at Hither Green being taken by 797-800. The merry-go-round continued to revolve.

Perhaps this continued movement of the King Arthurs was designed to confuse the enemy. If so, the strategy appeared to be successful, as – despite their

proximity to hostile forces – no King Arthurs were lost to enemy action. It was a near thing, though. Several N15s and King Arthurs were close by on Nine Elms shed when it suffered a direct hit from the bomb which destroyed T14 No.458. 751, 755, 775 and 776 were nearby and also N15X 2328, all of which received superficial damage. 806 nearly ran into a crater in July 1944.

At the end of the war thousands of civilians surged to the seaside and Waterloo struggled with the horde of passengers wanting to go to Bournemouth. By the second August Saturday, 764, 769, 778, 781 and 783 had been hastily summoned from the Eastern Section, together with several coach sets. The break-out of peace led to another King Arthur perambulation, resulting in the

immediate transfer of 767-769 to
Dover to bring troops home. Then,
after a more considered analysis
of post-war needs, taking into
account the availability of Bulleid's
new Merchant Navy pacifics, and
the first of the light pacifics, the
allocation was:

Nine Elms: 766, 775, 777
Salisbury: 448-457 plus 773, 774,
784 and 785
Exmouth Junction: 787, 789-792
Bournemouth: 772
Stewarts Lane: 763-765, 770, 771,
776, 778-783, 786, 788, 793-799,
800
Ashford: 801-806
Dover: 767-769

Although the new Bulleid Pacifics
took over the heaviest and fastest
expresses after the war (because of
lack of funds for electrification after

789 *Sir Guy* stands at Bournemouth Central with an up Sunday evening express, 1947.
(Sidney Boocock)

784 *Sir Nerovens*, outshopped in November 1945 still in wartime black livery, and transferred for a few months to Salisbury shed before returning once more to the Bournemouth line, on a West of England express, c1946.
(MLS Collection/E. Treacy)

the war, Bulleid was allowed to re-equip the Southern with his steam Pacifics) the King Arthurs were still much in evidence on weekday secondary passenger services and came out in force on peak summer Saturdays right through to their demise in the early 1960s.

Ossie Nock described one run on a Maunsell King Arthur shortly after VE Day in 1945, with 772 *Sir Percivale* on a Bournemouth train with Eastleigh men. The 6.20pm Sunday Waterloo was allowed a generous 100 minutes to Southampton with 420 tons, but checks were frequent – 772 took over half an hour to pass Walton just seventeen miles out, after several signal checks. Nock comments that 772 was a strong engine, the

evidence being its acceleration after the checks. Normally drivers notched up to 18 or 20% cut-off with full throttle, but this driver worked on 35% cut-off adjusting the regulator and seemed to go just as well. It maintained a steady 50mph up to milepost 31 and had just reached 60 mph after Fleet when signals at Winchfield intervened again. They were nearly eight minutes late past Basingstoke, but a good climb to Wootton Box, falling from 58 to 46mph, and 76½ down the bank before Eastleigh should have seen the train on time at Southampton, apart from yet another signal check at Eastleigh making the train 2½ minutes late in. Nock calculated the net time to be 87½ minutes for the 79.3 miles. They were just getting

back to pre-war standards of cleanliness and maintenance when nationalisation occurred.

And so the complete class of fifty-four Eastleigh, Scotch and Six-wheel tender Arthurs passed to British Railways on the 1 January 1948 to continue their valuable service for another ten to fourteen years.

9.4: The N15Xs

As described earlier in Chapter 8.4, much was initially expected of the rebuilt tender versions of the LB&SCR Baltic tanks and they were used at first on prestige trains, especially on the Bournemouth line. They made occasional appearances at Exeter, but stayed mainly east of Salisbury. It is possible that the designation

2333 *Remembrance* at Bournemouth Central awaiting departure with an up express for Waterloo, c1937. (John Scott-Morgan Collection/Rail Archive Stephenson)

N15X misled the loco depot staff and drivers into thinking they were improved versions of the King Arthur N15s, but they soon learned otherwise, for Maunsell had not taken the opportunity of incorporating a more modern front end with which he'd equipped the later King Arthurs.

Some South Western enginemen liked the new locomotives, but on the whole most did not take to them. They were soon relegated to semi-fast services to Salisbury, Basingstoke and Southampton and slow trains through to Weymouth. I searched through records from the *Railway Magazine* in the 1930s, and I found just a single reference and log which I show below, together with a comparison Cecil J. Allen's correspondent sent in of a Urie N15 on a similar load and schedule. The log appeared in the May 1935 edition of the *Railway Magazine* (see right).

The Remembrance run was respectable and good from Basingstoke to Wimbledon, but could not stand comparison with the excellent Urie N15. Cecil J. Allen quoted the South Western Section drivers as early as this saying that they were not enthusiastic about the new rebuilt Baltic tanks.

With the onset of war, they took their place on troop trains and freight services to Southampton Docks, and when the GWR required a loan of freight engines in 1941, the Southern Railway sent ten 4-6-0s, intending these to include six of the seven Remembrances, as they were universally known. The only one intended to be excluded was 2333

Salisbury – Waterloo, 1935			
	2329 *Stephenson* N15X 9 coaches, 280/295 tons		Comparison with 750 *Morgan le Fay* 10 coaches, 290/310 tons
Location	Mins Secs	Speed	Speed
Salisbury	00.00 (Comment – 2329 got away faster from rest)		
Tunnel Junction	03.20		
Porton	09.50		43
Grateley	17.20		36
Andover Junction	23.10	71	75
MP 62½	27.13	46	58½
Hurstbourne	29.00	55	65
Overton	35.20	49	62
Worting Junction	40.54	64	70
Basingstoke	43.05	66	74
Hook	47.50	75	77
Fleet	-	65	72
Farnborough	55.55	68	77
MP 31	57.58	61½	71½
Woking	63.40	75	76
Weybridge	68.20	55	62½
Surbiton	75.30	63	68
Wimbledon	81.05	sigs	
Clapham Junction	85.00		
Vauxhall	88.40		
Waterloo	91.45 (91 net)		

Remembrance herself, the Southern's war memorial engine. However, 2330 *Cudworth* was in Eastleigh Works under heavy repair at the time, so H15 478 was sent in its place and 2327-2329 and 2331-2 were joined by Urie S15s 496-499. 2327 and 2328 were allocated by the GWR to Old Oak Common; 2329 went to Newton Abbot; 2331 to Swindon and 2332 to Gloucester.

Their duties were almost invariably the wartime freight traffic and they returned to the Southern in July 1943 when the GW had built sufficient of the LMS 8F 2-8-0s and the American S160 2-8-0s were available on loan before they were required on the continent after the invasion of Normandy.

Unexpectedly, I found just one recorded wartime passenger train log with a Remembrance, the one that was in Works when the others were loaned to the LNER. As the log has no date, merely recording in the year of 1941, I assume it occurred before the Works visit and therefore it had run up a fairly high mileage at the time. The train was an up Exeter–Waterloo train, stopping at Andover and Woking, with 2330 coming on at Salisbury. Given war conditions and the general reputation of the N15Xs, the run was surprisingly sprightly.

On their return to the Southern all seven N15Xs were allocated to Nine Elms, but they were displaced by a new series of Bulleid Light

Salisbury – Woking, 1940 or 1941
2330 *Cudworth*
10 coaches, 322 / 345 tons

Location	Mins Secs	Speed
Salisbury	00.00	
Tunnel Junction	03.48	
Porton	10.18	42/38
Grateley	17.46	72
Andover Junction	24.08	
	00.00	
MP 62½	-	38
Hurstbourne	09.12	53
Whitchurch	11.24	50
Overton	15.19	60
Oakley	18.32	66
Worting Junction	20.29	
Basingstoke	22.34	76
Hook	27.07	66
Winchfield	29.12	
Fleet	32.07	69
Farnborough	35.06	
MP 31	37.14	58
Brookwood	40.04	65 / sigs
Woking	48.46	(44 net)

2331 *Beattie* roars past Surbiton with a Waterloo–Salisbury semi-fast service, c1947. (ColourRail)

Pacifics in April 1947 and the whole class was transferred to Basingstoke where they performed on commuter services to and from London and three or four-coach semi-fast and stopping services to Salisbury and Waterloo. They worked on the services from Reading to Portsmouth and in the summer they could also be found on cross-country services from the north via Oxford to Bournemouth, especially those that changed with the GW engine at Basingstoke.

Initially they were painted in the standard Maunsell olive green livery, which in common with many other locomotives, became black in the war years with sunshine yellow lettering. After the war they received the Bulleid malachite green livery with the yellow lettering on cabside and tender.

Despite their limitations, the class of seven was still intact at nationalisation and passed into the ownership of the Southern Region of British Railways.

9.5: The S15s

The Urie S15s soon settled into efficient heavy goods work, although the shortage of express locomotives in the early 1920s, before the King Arthurs appeared in sufficient numbers, meant that they were required to work some passenger services, especially relief holiday trains at the weekends, where some crews preferred them to the H15s. A decision was made to build more S15s incorporating Maunsell's improvements in

May 1926, and in the meantime the allocation of the Urie S15s in February 1927 was:

Nine Elms: 497-499, 501-504.
Feltham: 500, 505, 508-512
Salisbury: 496, 515
Exmouth Junction: 506

Although most of their work was east of Salisbury, the Salisbury, Exmouth Junction and Nine Elms engines were all seen in the Exeter area during the summers of 1927-29. However, once Maunsell's 823-832 series were run in, it became increasingly rare to observe any of the Urie S15s west of Salisbury.

The first ten Maunsell engines entered traffic in 1927, initially painted lined black, but were soon repainted green as they were so often used on passenger trains. They looked indistinguishable from

Urie S15 502, with a plain stovepipe chimney, rounds Clapham Junction curve with a down Portsmouth train, 27 August 1932. (J.M. Bentley/H.C. Casserley)

9.44pm Goods Templecombe–Salisbury
826, 44 goods vans, 500 tons
Driver Hayman, Fireman Sprag, Exmouth Junction

Location	Mins Secs	Speed	Mileage
Templecombe	00.00		0.0
Milepost 110	03.35	48	2.0
Milepost 107½	07.38	31	4.5
Gillingham	10.55	52½	6.8
Milepost 103¼	14.16	29	8.9
Semley	18.55	24½	10.9
Tisbury	26.00	-	15.9
Dinton	31.55	49	20.2
Wilton	40.30	-	25.9
Salisbury	47.05		28.4

the King Arthurs, apart from their 5ft 7in driving wheels, and soon showed that they were well capable of running at 70mph without any harm. 823-827 went initially to Exmouth Junction and 828-832 to Salisbury, and immediately started passenger work between Salisbury and Exeter where they quickly developed an excellent reputation. A further five, 833-837, built in 1928, were allocated to Feltham. Until 1937 they were often used on Portsmouth line passenger trains alongside the Schools, and were perceived as their equal, their better hill climbing abilities compensating for the Schools' faster downhill speeds.

Around this time, O.S. Nock experienced a footplate trip on Exmouth Junction's 826 on an evening fully fitted freight from Exeter to Salisbury, where the S15 was replaced by a King Arthur for the continuation to Nine Elms (see 9.3). As far as Templecombe, the load was a mere 200 tons, no test for the engine, but the load was made up there to forty-four wagons, an estimated 500 tons. Since the vehicles were all four-wheelers, resistance of the train would have made it considerably harder than a similarly loaded passenger train (Nock estimated it would have been equivalent to around 600 tons of passenger vehicles). He recorded the following train log (see above).

Nock commented that the climb to Semley required three-fifths

498 working empty carriage stock at Clapham Junction, 14 July 1935. (J.M. Bentley/H.C. Casserley)

Maunsell S15 825 on a freight service shortly after construction, c1928.
(Colin Garratt Collection/Rev A.W. Mace)

regulator and 35% cut-off, but halfway up the bank, the driver gave the engine a 'bit more' and held the speed at 24½mph steadily over the final mile. Cut-off was then reduced to 20% and the engine accelerated the train to 49mph under easy steam.

Prior to electrification of the Portsmouth line, the Urie S15s also frequently worked London–Portsmouth semi-fast passenger trains and even ocean liner specials to and from Southampton Docks, capable of holding 60-65mph and able therefore to keep time. They were robust engines and drivers found they could thrash them (as long as they were steaming well) without incurring hot boxes or other mechanical faults. They also had freight work over the Portsmouth line, including a very heavy regular turn from Nine Elms Yard to Eastleigh that frequently required the addition of a Drummond mixed traffic L11 or L12 4-4-0.

Like so many other Southern engines at this time, they were the subject of tender 'swaps', 504-510 losing their Urie bogie tenders to the N15X rebuilds in 1935, but regaining them from a variety of engines when they were withdrawn in the mid-1950s. In between, they received 4,000 gallon tenders off withdrawn Drummond 4-4-0s. The Maunsell engines also got involved, swapping their flush-sided eight wheel tenders with six wheel tenders from the Lord Nelsons, before receiving bogie tenders from Eastern Section King Arthurs.

827 emerges from the tunnel at Honiton West with a local passenger service for Exeter, 4 August 1928.
(J.M. Bentley/H.C. Casserley)

Maunsell S15 828 stops at Weybridge on a down Basingstoke local service, c1928.
(J.M. Bentley/Real Photographs)

Maunsell organised tests to improve the steaming of the Urie S15s in 1931 and 510 was fitted at Eastleigh with a modified blastpipe and King Arthur chimney in place of its Urie stovepipe. Steaming improved, but the short chimney was seen as unnecessary and a U1 Mogul pattern chimney was substituted on the low pitched boiler. This pattern kept the exhaust clear of the driver's cab, although the S15s were being fitted with smoke deflectors by this time. The remaining S15s received the modified blastpipe and chimney as they went through works. There was interchange between Urie and Maunsell boilers, although some drivers preferred the Urie boiler for goods work.

The Brighton line electrification in 1933 did not include freight services as the third-rail system was too dangerous for Yard shunting and marshalling. Maunsell's 833-837 were allocated for post-electrification freight work to the Central Section and were provided with short wheel-base six wheel tenders from King Arthurs 766-770. The acceleration of the Maunsell engines was needed to keep the freight trains clear of the new electric passenger trains.

One might have expected the Urie S15s to have been excluded from fast passenger work after delivery of all the Maunsell 4-6-0s and the 1937 Portsmouth route electrification, but Urie S15s were still being noted out in force on holiday weekends right up to 1939, especially if the seasonal peak coincided with ocean liner arrivals or departures. In July 1939, on four successive Saturdays, between fifteen and eighteen Urie S15s were noted passing Basingstoke on passenger trains. One S15 was working a Lymington Pier–Waterloo train, most were on holiday reliefs or Southampton Docks boat trains. Once war broke out they were needed for freight and troop trains.

831 on a Salisbury semi-fast train rushes past
Walton-on-Thames, 29 October 1938.
(J.M. Bentley/ H.C. Casserley)

Ten more S15s were built by
Maunsell in 1936 – 838-847 – and
the first five were immediately sent
to Hither Green for the Eastern
Section to take over the heaviest
freights from the 'N' Moguls. 843-
846 augmented the fleet at Feltham,
and the last one, 847, was allocated
to Exmouth Junction. On the Eastern
Section, the new engines were
restrained to a maximum of 45mph
and forbidden to haul passenger
trains. I can find no explanation
for this other than concerns about
the quality of the track or the
extreme conservatism of the local
management. They had plenty of
demanding work however, with
four main line goods turns to
Ashford and Dover via Tonbridge.
A fifth engine, when not required
to cover any under maintenance,
would undertake cross-London trip
working. Despite the management's
caution, the Maunsell S15s were very
popular with the Eastern Section

crews and depot staff who kept
them in excellent external condition.
They were masters of the climb to
Hildenborough after the Tonbridge
slack and would outclass a Mogul on
this stretch. However, at the outbreak
of war, they were all transferred
to Feltham, five King Arthurs
successfully taking their place.

The five Brighton-based S15s were
also reallocated at the start of the
war – to London, but to New Cross,
working both freight and troop
specials. Then in 1943 they too joined
many of the others at Feltham, now
the home of most of the class. The
engines were robust and simple,
and in the wartime conditions were
usefully averaging 131,000 miles
between Works repairs.

At the onset of the war, the
whole class of Urie S15s was also
reallocated to Feltham apart from
496-499 which were part of the
group of Southern locomotives
loaned to the Great Western from

1941 until the spring of 1943.
Although the GW crews preferred
their own Swindon products, the
four Urie S15s were well received,
certainly better than the N15Xs
loaned at the same time. 496 was
initially based at Southall, 497-499
at Old Oak Common, but 496 and
497 moved later to Tyseley. The
engines had clearly been worked
extremely hard during their exile,
for they returned in a very run-
down condition and had to be given
heavy overhauls at Eastleigh Works
before they could resume activity
on the Southern.

In February 1944, Feltham
received ten new WD 2-8-0s
which released many of the Urie
S15s, which had been worked to
exhaustion, to undergo heavy
overhauls at Eastleigh. When the
WDs were required for service in
France and the Netherlands, the
S15s were back in traffic in good
condition.

In December 1946, at the end
of the hostilities, there was a
redistribution of the class to meet
peacetime freight needs and the
Maunsell S15s were reallocated as
follows:

Feltham: 833-843
Salisbury: 828-832
Exmouth Junction: 823-827, 844-847

Freight traffic had decreased on the
continental routes, so vital during
the war, and the power of these
engines was needed on the Devon
and Wiltshire banks as fast freight
traffic of perishables grew back
again.

During the war both Urie and
Maunsell S15s were painted black
and all passed at nationalisation
to the Southern Region of British
Railways.

Chapter 10

URIE AND MAUNSELL ENGINES IN BRITISH RAILWAYS OPERATIONS

10.1: The H15s

The first evidence of nationalisation for the H15 class was the outward appearance, the class being classified as 'mixed traffic' and therefore entitled to the BR standard livery of lined black. 30484 and 30490 received their new numbers and livery in November 1948 and the rest of the class soon followed, with the exception of 1924-built 30477 and the N15 boilered 30491, which appeared to have been painted in the goods livery of unlined black in error, which was not corrected until their next major overhauls in November 1952 and September 1950 respectively.

Immediately after nationalisation the allocation was split between just three of the new Southern Region depots:

Nine Elms: 30477, 30482-30491
Eastleigh: 30473, 30474, 30477, 30521-30524
Salisbury: 30330-30335, 30475, 30476

In the early days of nationalisation there was still a shortage of express locomotives at peak holiday periods despite the influx of large numbers of the Bulleid Light Pacifics, and H15s found themselves deputising, sometimes on the most surprising of turns. Salisbury's 30330, 30332 and 30333 were noted on the up *Atlantic Coast Express* in the summer of 1950 and apparently restricted loss of time on the Exeter–Salisbury leg to less than ten minutes, despite the usual thirteen-coach load, although post-war schedules were still lax.

F13 rebuild H15, 30331, heads a freight at Hinton Admiral station, c1952. (MLS Collection)

Salisbury H15 30334 enters its home station with a stopping train from Yeovil Junction, c1957. (John Hodge)

Rebuilt F13 30332 on Eastleigh shed, 5 May 1955. (Colin Boocock)

The Nine Elms engines were used on reliefs to Bournemouth and ocean liner specials and were joined for a short while by 30334 and 30335 from Salisbury, which were rostered for a time on the 7.48 and 9.54am Waterloo-Bournemouth.

By 1952, such appearances were rare and usually only in loco failure substitutions such as 30483 replacing SR main line diesel 10201 on the 11.30am Weymouth–Waterloo in August. In 1953, the motive power situation had worsened again – the mainline diesels had been transferred to the LMR – and both H15s and S15s found themselves on express duty. The author experienced a non-stop run from Salisbury to Waterloo on a holiday train from Sidmouth Junction around this time, although I have to admit that 30523 did not distinguish itself. 30476 did rather better on the down *Royal Wessex* in July 1953, only losing eleven minutes – rather better than the same engine three years later when it was seen struggling with the 3.20pm Waterloo–Bournemouth, a Merchant Navy diagram.

30485 and 30490 were withdrawn in 1955, then over forty years old – 30485 after a collision with light engine 30783 *Sir Gillemere*, as it was arriving with a stopping train from Weymouth at Bournemouth Central. 30490's large bogie tender was retained and attached to Eastleigh Arthur 30457 *Sir Bedivere*, the only use of these tenders after their owners were scrapped. Despite this, other members of the class were still to be found on seasonal and Bank Holiday relief trains, including the 11.22am, 12.35pm,

A freight arriving from the Salisbury direction at Eastleigh, 3 July 1956. (Colin Boocock)

1.5pm and 2.34pm Waterloo to Salisbury or Bournemouth. There was a regular H15 or S15-hauled Wimbledon–Weymouth Saturday train and the 10.21am Sidmouth Junction–Waterloo from Salisbury, which was presumably where I experienced 30523 in 1954. The

The solitary E14 rebuild, 30335, ready to depart Andover Junction with a short eastbound freight. (ColourRail)

30487 heads a heavy up freight away from Eastleigh, 23 May 1955. (Colin Boocock)

30490 approaches Southampton Central with a stopping train from Bournemouth, April 1955. (Colin Boocock)

Eastleigh members of the class were frequently employed on ocean liner specials to and from Southampton Docks, especially if those coincided with heavy holiday period traffic.

Apart from freight and parcels trains, the H15s could regularly be seen on the Waterloo–Basingstoke passenger semi-fasts, replacing the Drummond T14s when they finally went in the early 1950s. A couple of the Salisbury F13 rebuilds, 30330 and 30332, were withdrawn in 1956-7, along with Nine Elms' 30483 and 30487, but 30331, 30486, 30489, 30491, 30521 and 30523-4 received heavy repairs during these years and survived until 1959-61, when engines released by the Kent Coast electrification were transferred to the Region's Western Section. General repairs ceased with 30522 in March 1959. 30477 and 30478 were condemned shortly afterwards in the first half of 1959. Eastleigh's remaining four, 30473-30476, often powered the empty van train from Eastleigh to Nine Elms, following the up ACE onto the main line at Hampton Court Junction. The Nine

The hybrid – Urie's 30491 with a taper boiler, on an Up train at Southampton Central, 11 October 1953.
(Colin Boocock)

30485, running in with a local train from Weymouth, collided with Scotch Arthur 30783 *Sir Gillemere* which had run light out of shed through the disc signals. The damage to 30485 was severe enough to cause its withdrawal without any repair work being justified, 23 January 1955. (Colin Boocock)

30490, the first Urie H15 to be withdrawn, arrives at Bournemouth with a stopping service from Weymouth, c1955. MLS Collection

of Nine Elms' four King Arthurs. The last two of the Urie 1914 build, 30489 and the taper-boilered 30491, regularly appeared on Salisbury or Basingstoke–London semi-fasts until the end of 1960, turn and turn about with King Arthurs or the Region's Standard 5s.

The final allocation before withdrawal of the class was:

Nine Elms: 30482-30491, 30521-30524
Eastleigh: 30473-30478
Salisbury: 30330-30335

The last booked diagrams for these engines included just two Salisbury duties – both Meldon Quarry ballast trains between Salisbury and Woking, being serviced at Guildford and returning with the empty ballast wagons, and an Eastleigh turn involving

Maunsell-constructed H15 30475 drifts through the New Forest with a freight for Bournemouth and Weymouth, c1957. (MLS Collection)

Elms four, 30521-30524, frequently worked the 11.15pm Waterloo last semi-fast to Basingstoke, returning on a freight, a turn booked for one

30478 approaches Salisbury station with a down semi-fast, c1957. (John Hodge)

The 3.28pm stopping train from Templecombe approaches Salisbury from the west hauled by Nine Elms H15 30521, 17 August 1957. (John Hodge)

30524 hurries up the slow line approaching Woking with a Southampton Docks– Nine Elms perishable freight, July 1959. (Author)

freight trips to Southampton Docks, an Eastleigh–Portsmouth slow passenger train and a late evening freight turn from Fratton to Chichester and back to Eastleigh. Their other use was substituting for King Arthur turns as instanced in the preceding paragraph.

The last survivors were all from the 1924 build under Maunsell, 30475, 30476, 30521 and 30523. All were condemned at the end of 1961. The H15s had had a long and unspectacular career performing useful work for some forty-five years, with 30489, withdrawn in January 1961, being credited with 1,521,178 miles and 30491, withdrawn a month later, with 1,539,740. 30521 was the highest of the 1924 locomotives, clocking up 1,161,139 miles. The F13 rebuilds were credited with even more (30332, 1,791,908) although this included their work as the Drummond unrebuilt engines, little of which survived the conversion. 30335 ran 1,327,650 as an H15 and 968,484 miles of 30332's total mileage was as the rebuilt engine. The last reported working of an H15 was 30523, sighted at Salisbury on a van train on the 6 December 1961.

The legacy of the Urie H15s lived on for a while, for they were Britain's first 2-cylinder 4-6-0s with Walschaerts valve gear, developed over the years as the 'Black 5s' on the LMS, the LNER's Thompson B1s and finally the BR Standard 5 4-6-0s, some of which ended their time on the Western Section of the Southern Region working the services that a year or two earlier had been the preserve of the H15s.

30523 sails over the Worting Junction flyover with an Eastleigh–Nine Elms ECS and van train, c1955.
(J.M. Bentley Collection)

Breakdown cranes attend the derailment of 30477 at the outlet points to Eastleigh depot, 8 February 1955.
(Colin Boocock)

30521 leaves Eastleigh with a freight for the Salisbury line, 3 July 1956. (Colin Boocock)

30522 leaves Eastleigh with a Bournemouth – Waterloo stopping service, 3 July 1956. (Colin Boocock)

30473 waits to back down into Southampton Terminus station, with the Docks in the background, 26 June 1957. (R.C. Riley)

30524 drifts down the bank towards Dorchester after leaving Bincombe Tunnel with a Bournemouth–Weymouth stopping train, c1959. (ColourRail)

The Urie H15 with the Maunsell taper boiler, 30491, pausing at Hinton Admiral station with a three-coach Bournemouth–Southampton Central stopping train, 27 February 1960. (R.C. Riley)

30486, the first Urie H15 built in December 1913, heads a three-coach local train from Eastleigh to Bournemouth, c1959.
(John Scott-Morgan Collection)

30489 on arrival at Waterloo with the first up semi-fast train, the 6.39am Basingstoke, 1958. 30489 was the last survivor of the parallel boiler Urie H15s, being withdrawn in January 1961 after running 1,521,178 miles since its entry into traffic in May 1914.
(R.K. Blencowe Collection/ P. Winding)

10.2 The N15s

All twenty Urie N15s entered the British Railways stock list in active mode on 1 January 1948 and were progressively renumbered 30736-30755 through 1948 and 1949 to May 1950 when, finally, 30747 got its nationalised identity. 'British Railways' replaced 'Southern' on their tenders although they initially retained their malachite green liveries. The standard BR passenger livery of the darker Brunswick green was first applied to 30741 *Joyous Gard* in September 1949. The last two to retain malachite green livery were 30736 *Excalibur* and 30752 *Linette*, both wide chimney multiple jet-blastpipe engines.

The entire class seemed still in comparatively good mechanical

30736 *Excalibur*, with multiple jet exhaust, doubleheads a class 'U' Mogul at Parkstone on a Weymouth–Bournemouth local service made up of Eastern Region rolling stock, 1953. (MLS Collection)

30753 *Melisande* passes Durnsford Road Power Station on a Waterloo–Basingstoke semi-fast service, May 1953. (Transport Treasury/R.C. Riley)

30744 *Maid of Astolat* on a Waterloo – Basingstoke semi-fast train at Waterloo alongside Doncaster-based V2 60896, loaned to the Southern Region during the temporary withdrawal of the Merchant Navy Pacifics after an axle fracture at speed, heading the 12.30pm down *Bournemouth Belle* Pullman train, 21 May 1953. (Transport Treasury/R.C. Riley)

30750 *Morgan le Fay* near Brockenhurst on a relief Waterloo–Bournemouth express, 24 May 1953.
(Transport Treasury/R.C. Riley)

condition and although they were kept smart, several – 30738, 30740-30743, 30751, 30752, 30754 and 30755 – spent 1949 and part of 1950 in store. The others remained well occupied, on semi-fast passenger services to Basingstoke, Salisbury and Bournemouth on weekdays and from Easter 1950, all came out in force on holiday reliefs. Both 30744 and 30747 were seen on West of England trains through to Sidmouth Junction or Exeter. For a couple of years one N15 was based at Feltham for a regular Feltham Yard–Reading goods train, initially 30738, then 30744 after which the latter engine was stored until transferred to Nine Elms for the Christmas 1951 traffic. The freight turn subsequently reverted to Nine Elms and no further N15s were allocated to Feltham until a few of the Maunsell engines went there after the Kent Coast electrification.

The first casualty was 30754 *The Green Knight*, which was withdrawn from Basingstoke shed in January 1953 as a result of a fracture in its main frame, with a mileage of 1,151,284. The others soldiered on with much work that year, including heavy West of England relief trains which sometimes they worked right through to Exeter, merely changing crews and taking water at Salisbury. That year, 1953, the allocation was:

Nine Elms: 30744, 30750-30752, 30755
Basingstoke: 30745, 30749, 30753
Eastleigh: 30746-30748
Bournemouth: 30736-30743

The Bournemouth engines, apart from local services to Weymouth or Southampton, mainly worked the heavy cross-country services

30747 *Elaine* **on** a West Midlands–Bournemouth cross-country train near Lymington Junction, 1 August 1953. (Transport Treasury/R.C. Riley)

from Poole and Bournemouth to Basingstoke, Reading or Oxford, where they would change engines to Western Region power before the trains went forward to Wolverhampton, Birkenhead, Sheffield or Newcastle, returning on the equivalent southbound trains.

The RCTS undertook a survey on 22 August 1953 and the following N15 use was noted:

Nine Elms engines:
30744 2pm Waterloo–Southampton Ocean Liner Special and 5.05pm return
30750 10.05am Waterloo–Bournemouth Central
30751 7.52am Waterloo–Bournemouth West and 12.30pm return
30752 11.22am Waterloo–Weymouth and 7.20pm return
30755 under repair at Nine Elms
Basingstoke engines:
30745 6.37am Basingstoke–

Waterloo and 1.5pm Waterloo–Exeter relief
30749 Up side pilot (not utilised)
30753 Basingstoke–Waterloo semi-fast services.
Eastleigh engines:
30747 Channel Island boat train.
30748 9.2am Southampton Central–Waterloo, 12.50pm Waterloo–Bournemouth Central.
Bournemouth engines:
30736-30738 Bournemouth cross-country trains to Oxford
30739 10am Bournemouth–Waterloo and 2.34pm return
30740 10.8am Bournemouth Central–Waterloo
30741 10.20am Bournemouth West–Waterloo and 3.20pm Waterloo–Weymouth
30742 12.20pm Weymouth–Waterloo.
30743 6.35am Bournemouth West–Waterloo stopping train
30746 Bournemouth cross-country train to Oxford.

30739 *King Leodegrance* on a Weymouth–Bournemouth local train arrives at Dorchester South, backs into the main platform and departs, 27 August 1954. (Transport Treasury/R.C. Riley)

It is noteworthy that out of the nineteen extant engines of this class, eighteen were working that day and only 30755 was on shed under repair. Of locomotives allocated to the Southern's Western Section and not in Works at the time, the Urie N15s scored the highest availability (95%) followed by the Lord Nelsons (94%), the Merchant Navies (91%) and the Maunsell King Arthurs (90%). Only the Bulleid Light Pacifics at 77% were appreciably lower.

In 1954 the Urie N15s were still active on important expresses as well as their usual diet of semi-fast and stopping services around Basingstoke and Bournemouth. The Bournemouth engines in particular seemed active, with 30740 making an appearance, presumably after a failure by a Bulleid Pacific, on the down *Royal Wessex*, while 30738 *King Pellinore* and 30739 *King Leodegrance* were regularly diagrammed for a time on the 6.30pm Waterloo–Bournemouth, which was first stop Southampton.

Further withdrawals took place in 1955 with the loss of 30740, 30743, 30746 and 30752, the latter the first withdrawal of an N15 modified by Bulleid with wide diameter chimney and multiple jet blastpipe. 30740 *Merlin* was reprieved for a few months in order to bow out in a spectacular way, being scheduled to be blown up on live television. The deed was to take place on the Longmoor Military Railway where a rail in front of the engine had been removed and a bomb detonated as the driverless engine moved towards its doom with the excited voice of the commentator getting

30740 *Merlin* **on** the down relief line beside the River Thames at Tilehurst with a Bournemouth–Birkenhead cross-country express, 17 September 1955. (Transport Treasury/R.C. Riley)

ever more shrill. There was finally an anti-climactic 'thud' and *Merlin* slid gracefully down the bank still upright and blowing off steam. It was so little damaged by this experience that it was able to be towed on rail to Brighton Works in May 1956 for breaking up. I believe this was yet a further demonstration of the robustness with which the Urie engines were built – unfortunately balancing this with their heavy weight for the power produced and the hammer blow

Wide-chimneyed multiple jet exhaust Urie Arthur, 30736 *Excalibur*, passes Tilehurst station with an up cross-country express from the West Midlands to Bournemouth, 17 September 1955. (Transport Treasury/R.C. Riley)

30742 *Camelot* **sweeps** round the curve on the relief line at Pangbourne with an up cross-country express for Bournemouth, 17 September 1955. (Transport Treasury/R.C. Riley)

they exerted on the track. However, in contrast to *Merlin*'s fate, 30738, 30748, 30750, 30751 and 30755 were given heavy overhauls during the year and fourteen of the class played a significant role in the 1955 summer season's holiday traffic.

In 1956, the Southern Region had its full allocated fleet of BR Standard 5 4-6-0s, 73080-73089 going to Stewarts Lane and 73110-73119 to Nine Elms. 30750 *Morgan le Fay* was then the only Urie N15 left at Nine Elms. This engine was still very active during the summer and the author noted it at Woking on 28 July (a summer Saturday) working the 10.58am Exeter–

30748 *Vivien* **near** Farnborough with an evening Waterloo–Basingstoke commuter train, probably the 5.09pm Waterloo, c1955. (ColourRail)

Waterloo, looking in good shape and blowing off steam as it was checked by signals approaching the station. However, six N15s were condemned that year, 30736, 30737, 30741, 30744, 30745 and 30747, the latter engine also having been seen active on 28 July that year on an Up Lymington Pier express – presumably after a failure as 4-6-0s could not be accommodated on the turntable used for the Lymington Pier train engines. 30753 was also sent to Brighton for breaking up in March 1956, but was given a light repair and returned to work at Basingstoke after returning home part-way on a Brighton–Redhill passenger working, a rare occasion for an engine with a Urie cab to work over that route.

Charterhouse School's Railway Society undertook a survey of traffic between Weybridge and Basingstoke from 12noon until the early evening on 28 July 1956, and noted six Urie N15s and eighteen Maunsell King Arthurs in operation. In fact, the survey was remarkable for the fact that the Urie and Maunsell King Arthurs far outnumbered any other class of locomotive at work on the South Western Section of the Southern that very busy day (often known as 'Black Saturday', the first Saturday after the break-up of schools with many families taking their fortnight's holiday on the South Coast). For full details of that survey, see chapter 11, when

Basingstoke's 30753 *Melisande* on a morning Basingstoke–Waterloo commuter train, near Brookwood, c1955. (ColourRail)

30749 *Iseult* **backing** out of Waterloo after arriving with an inbound train from Basingstoke, 7 September 1954. The generator for the electric lighting is very visible on the running plate just behind the smoke deflector. (Transport Treasury/R.C. Riley)

30749 *Iseult* approaches Surbiton with an evening commuter train for Basingstoke, probably the 5.9pm from Waterloo, c1953. (Robin Russell)

30749 *Iseult* again, departing friom Basingstoke with a Birkenhead – Bournemouth express which the Urie N15 would have taken over from a Western engine at Oxford, 27 October 1956. (Transport Treasury/R.C. Riley)

The last Urie N15 survivor, 30738 *King Pellinore*, approaches Weybridge station at speed with a morning semi-fast train from Basingstoke to Waterloo, January 1958. (H.C. Casserley)

unfortunately it will be revealed that despite the presence of the N15s on so many workings, two of them let the side down badly and caused bottlenecks and threw the day's timekeeping on nearly all services out of the window!

At the beginning of 1957 the allocation was:

Nine Elms: 30748, 30750
Basingstoke: 30749, 30751, 30753, 30755
Eastleigh: 30738
Bournemouth: 30739, 30742

As withdrawals continued, the remaining engines congregated at Basingstoke where their heaviest regular duties were ten-coach commuter services to and from Waterloo, and the occasional Ocean Liner Express, replacing all that

depot's N15X Remembrances. Otherwise they ran middle of the day and evening semi-fast services to London, stopping trains to Salisbury, van trains from Eastleigh to Nine Elms and Waterloo and freight trains to Woking or Feltham. Most of the remaining N15s were withdrawn in 1957. The last survivors were 30748 *Vivien* and 30738 *King Pellinore*, the latter engine being noted in the summer of 1957 on up and down boat train specials, and during the autumn often reached Waterloo on the 7.10am from Yeovil.

King Pellinore was finally withdrawn in March 1958, having achieved a mileage of 1,460,218 in its forty-year life. Only two N15s exceeded this – 30744 *Maid of Astolat* with 1,463,292 and 30745 *Tintagel* on 1,464,032. A dozen were scrapped at Eastleigh, the other eight at Brighton Works. Their ghosts continued to haunt the South Western main line for

another ten years in the form of the BR Standard 5s, which acquired all twenty names of the Urie N15s and even had nameplates that were shaped like those of their proper predecessors.

10.3 The King Arthurs

By the time of nationalisation, the Southern Region of British Railways owned a fleet of thirty Merchant Navies and around ninety light pacifics, so it would not have been surprising if the N15s and King Arthurs had totally disappeared from express passenger duties. However, the availability of the new Bulleid West Country and Battle of Britain engines was not all it might have been, and the King Arthurs still stepped in, and during the holiday season were prevalent on all the non-electrified main lines – to the Kent resorts via both Tonbridge and Chatham and to Bournemouth, Salisbury, and on summer Saturdays to Exeter.

All fifty-four King Arthurs were intact at nationalisation and, like the other Maunsell and Urie engines, had 30,000 added to their SR numbers, becoming 30448-30457 and 30763-30806. As with the Urie engines, they continued to be painted malachite green with their new numbers on the cabside and British Railways on the tender until July 1949, when 30783 appeared in the adopted brunswick green eventually chosen for all but the most powerful passenger locomotives. 30786 and 30801 were the last engines to get a malachite green repaint in the same month.

With so many Light Pacifics then available and still more emerging

Scotch Arthur 30772 *Sir Percivale* enters Oxford with a Bournemouth–West Midlands cross-country express as a Worcester Castle, believed to be 4086 *Builth Castle*, awaits departure for Paddington, c1949.
(Colin Garratt Collection/Rev A.W. Mace)

30785 *Sir Mador de la Porte*, with a motley collection of rolling stock, heads a Bournemouth – Eastleigh stopping train from Christchurch, c1948.
(J.M. Bentley Collection)

new from Eastleigh Works every month, a reallocation of King Arthurs was inevitable. The last King Arthurs had already lost all their work at Exmouth Junction before nationalisation and the only ones seen west of Salisbury on a regular basis afterwards were the Salisbury engines, mainly on slow passenger or freight trains, except in the height of the summer season. In June 1949, they were distributed as follows:

Nine Elms: 30773, 30780, 30782, 30783, 30786, 30791, 30792
Salisbury: 30448-30457
Eastleigh: 30777, 30779, 30784, 30785, 30789, 30790
Bournemouth: 30772, 30787, 30788
Stewarts Lane: 30763-30766, 30774-30776, 30778, 30793-30797
Hither Green: 30800
Bricklayers Arms: 30798, 30799
Dover: 30767-30771, 30781
Ashford: 30801-30806

The lone Hither Green example worked the 5.40pm Cannon Street–Deal, returning on an overnight freight, 12.50am Ashford–Hither Green, and, as the only passenger engine at that depot, was cosseted and kept in immaculate condition. Later, to their chagrin, the shed staff lost their star, but it was replaced by 30806 *Sir Galleron* which then adopted the 'star' mantle.

In 1950, according to a contemporary *Trains Illustrated*, 142 Southern Region steam engines were put into store at the beginning of the winter service, including twenty-nine N15s and King Arthurs. The superabundance of power was not met by withdrawals, however, and most of the King Arthurs were taken out of store for the Christmas postal

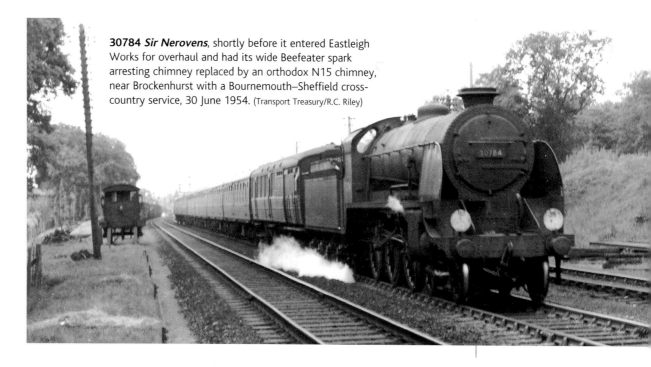

30784 *Sir Nerovens*, shortly before it entered Eastleigh Works for overhaul and had its wide Beefeater spark arresting chimney replaced by an orthodox N15 chimney, near Brockenhurst with a Bournemouth–Sheffield cross-country service, 30 June 1954. (Transport Treasury/R.C. Riley)

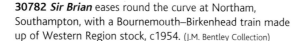

30782 *Sir Brian* eases round the curve at Northam, Southampton, with a Bournemouth–Birkenhead train made up of Western Region stock, c1954. (J.M. Bentley Collection)

30783 *Sir Gillemere* between Bournemouth and Southampton with an up Bournemouth – Waterloo summer Saturday train, c1955. (J.M. Bentley Collection)

Bournemouth's 30783 *Sir Gillemere* waits to depart Waterloo with a holiday express for Bournemouth while M7 30241 of Nine Elms exits the adjacent platform after bringing in the empty stock and banking a previous train out of the platform, c1953. (J.M. Bentley Collection)

traffic. Timekeeping had deteriorated on the South Eastern Section, whether the motive power was an Arthur, Schools or Light Pacific and those King Arthurs with six-wheel tenders suffered, along with the Schools, in needing to take water more often. With better coal and a generally tightening up of performance, things had got back to reasonable normality by 1952-3 and the N15s and King Arthurs were out again in force as the 1950s holiday season built up to levels not experienced before.

A large number of transfers took place in 1951 which appears to have been for administrative rather than traffic purposes, grouping engines of the same class in numerical sequence. By the summer service, the King Arthurs were allocated thus:

Nine Elms: 30456, 30457, 30780, 30781
Salisbury: 30448-30455
Eastleigh: 30784-30790
Bournemouth: 30782, 30783
Stewarts Lane: 30763-30769, 30777-30779, 30791-30795
Hither Green: 30806
Bricklayers Arms: 30799-30801
Dover: 30770-30776, 30796-30798
Ashford: 30802-30805

These transfers caused resentment at some of the smaller depots where their well-maintained engines were replaced by more well-worn engines from larger depots.

In the early 1950s, O.S. Nock made a number of footplate trips on King Arthurs on the Eastern Section. One noted performer at this time was Stewarts Lane's 'pet' 30768 *Sir Balin*, kept in immaculate condition and regularly rostered to the 3.35pm Victoria–Margate. The driver was the celebrated Sam Gingell, with his regular fireman

Rowe. At the time of the run logged below 30768 had already amassed 70,000 miles since its last overhaul and went on to clock up over 90,000 miles before Works attention was necessary.

Nock drove the engine from Margate to Ramsgate using 40% cut-off and full regulator up the 1 in 80/110 to Broadstairs and the train arrived in Ramsgate a minute early. He comments that for an engine of such high mileage it ran remarkably smoothly, even when linked up to

30789 *Sir Guy* with a down freight approaching Basingstoke, c1955. (J.M. Bentley Collection)

30451 *Sir Lamorak* leans to the curve at Worting Junction with a Waterloo–Salisbury semi-fast service, c1956. It still retains its watercart tender which it will lose in January 1957. (J.M. Bentley Collection)

30771 *Sir Sagramore* approaches Eastleigh with a Sheffield–Bournemouth cross-country express, 7 September 1957. (John Hodge)

Bournemouth's 30782 *Sir Brian* takes over a Sheffield–Bournemouth cross-country train from a Western Region engine at Oxford, 29 September 1956. (R.C. Riley)

15%. The Arthur demonstrated its sure-footedness by never slipping and showing how suitable these engines were for medium and heavy stopping services, more so than either a Schools or Light Pacific. He returned on the engine again on an evening Ramsgate–Victoria train which left Chatham 2½ minutes late after station overtime. He published his log only for the Chatham–Bromley South section of line (see over page).

The train then followed another, getting signal checks all the way to Victoria.

In the summer of 1954 Nock had the opportunity of a further footplate run on a King Arthur on a heavy continental boat train to Dover Harbour. It was a relief to the regular 10am London-Ostend-Brussels service and was allowed 113 minutes for the 81 miles via Maidstone with its curvature and severe speed restrictions.

Nock estimated this performance required usage of 45 gallons of water per mile and around 60lb of coal. He calculated an indicated horsepower of 1,340 at Westenhanger, equivalent drawbar horsepower of 1,065 with similar power outputs at Bearsted and Ensden Tunnel. The tender was topped up with soft Kentish coal for the return journey and 30769 was rostered for the second of three boat trains connecting with the morning ferry from Ostend.

The line was clearly congested from Swanley Junction onwards and Nock ceased taking detailed times and speeds as they were clearly following a succession of trains. Net time was impossible to calculate.

3.35pm Victoria – Margate/Ramsgate
30768 *Sir Balin*
9 coaches, 301 / 315 tons

Location	Mins Secs	Speed	(Schedule)	Engine Working
Victoria	00.00			
Wandsworth Road	05.15			
Herne Hill	08.27	38	(8½)	
Sydenham Hill	11.37	28		
Beckenham Junction	15.28	60/pws 20*	(16½)	
Bromley South	19.45	(20)		
	00.00			
Bickley Junction	05.12		(5)	
St Mary Cray	07.55	65		
Swanley Junction	10.28	60	(11)	
Farningham Road	13.09	79		20% cut-off, full reg
Fawkham	15.27	69		
Meopham	17.58	58/62		
Sole Street	18.58	60	(22)	
Cuxton Road	-	79		15% cut-off. Full reg
Rochester Junction	26.20		(30½)	
Chatham	28.13		(33)	
	00.00			
Gillingham	04.00			
Newington	10.47	60/56/66		20% cut-off, full reg
Sittingbourne	14.30		(13)	
	00.00			
Milepost 46	02.47			
Teynham	04.57	56/51½/64½		
Faversham	09.47		(11)	
	00.00			
Graveney Siding	04.46	61		
Whitstable	08.52		(10)	
	00.00			
Chestfield	03.58		(4)	
	00.00			
Herne Bay	04.07		(5)	
	00.00			
MP 63½	02.56			
MP 64	04.11	30		40% cut-off, full reg
MP 69½	09.32	74		20% cut-off, full reg
Birchington	11.07		(12)	
	00.00			
Westgate	04.36		(4)	
	00.00			
Margate	03.36		(4)	

9.25pm Chatham–Bromley South/Victoria
30768 *Sir Balin*
9 coaches, 286/300 tons
Driver S.Gingell, Fireman F.Rowe

Location	Mins Secs	Speed	(Schedule)	Engine Working
Chatham	00.00			
Rochester Bridge Jcn	02.02		(2½)	Engine priming
Cuxton Road Box	07.25	35		45% cut-off, full reg.
Sole Street	15.00	31	(15)	
Meopham	16.25			
Fawkham	19.07			
Farningham Road	21.29	86		
Swanley Junction	23.55	60	(26)	
St Mary Cray	26.37	69		
Bickley Junction	30.57	sigs 5*	(32)	
Bromley South	34.25		(35)	

The 'Pride of Stewarts Lane', the immaculate 30768 *Sir Balin*, ready to leave its home depot to run up to Victoria ready to head its regular turn, the 3.35pm Victoria–Ramsgate, c1954. (J.M. Bentley Collection)

9.40am Relief Boat Train Victoria–Dover Marine, summer 1954
30769 *Sir Balan*
12 coaches, 370/405 tons
Driver T.Tutt, Fireman R.Wilkes, Stewarts Lane

Location	Mins Secs	Speed	(Schedule)	Engine Working
Victoria	00.00			
Wandsworth Road	04.30	34½		
Herne Hill	09.20	sigs 2*	(8½)	
Sydenham Hill	14.02	25		
Beckenham Junction	18.30	55		
Bromley South	23.32	sigs 20*		
Bickley Junction	28.00	sigs	(23)	
St Mary Cray	31.42	sigs		
Swanley Junction	36.05	sigs stop	(29)	
Eynsford	43.47	39/29		
Otford	51.40	44	(38)	
Kemsing	54.57	30½	(1 in 82)	35% cut-off, full reg.
Wrotham	58.51	53/pws 15*		
Malling	65.34	66		17% cut-off, full reg.
Barming	68.17	56		
Maidstone East	73.01	15*	(60 ½)	
Bearsted	79.06	30/18	(1 in 60)	40% cut-off, full reg.
Hollingbourne	82.59	39/28		
Harrietsham	87.13	40½		25% cut-off, full reg.
Lenham	90.30	28		35% cut-off, full reg.
Charing	95.05	61		17% cut-off, full reg.
Hothfield	97.32	69		Regulator eased
Ashford	101.44	slack	(88)	
Smeeth	106.45	55½		
Westenhanger	110.55	53	(1 in 286)	27% cut-off, full reg.
Folkestone Central	116.55	62	(104)	
Folkestone Junction	117.54	60		
Shakespeare Box	122.55	sig stop		
Dover Marine	129.45	(109 net)	(113)	

30767 *Sir Valence* on the down *Kentish Belle* Pullman, overtakes 'N' 31404 on a Victoria–Ramsgate express at Bickley Junction, 5 August 1956. (R.C. Riley)

30767 *Sir Valence* with a down Ramsgate express at Downs Bridge at the summit of the 1 in 100 between Beckenham Junction and Shortlands Junction, 4 April 1958. The photographer's garden backed onto the line at this point and Richard Riley was a frequent house guest and also took photos from this vantage point.
(Ken Wightman)

30795 *Sir Dinadan*, running in from repair at Eastleigh in October 1955, works a local train down the Bournemouth line. It was overhauled at Eastleigh and returned to the Eastern Section, received a bogie tender in April 1958 and was withdrawn from Basingstoke shed in August 1962.
(MLS Collection)

30791 *Sir Uwaine* with the *Kentish Belle* Pullman train, passing Bickley Junction, c1954. (MLS Collection)

2.50pm Dover Marine – Victoria Relief Continental Boat Train, summer 1954
30769 *Sir Balan*
12 coaches, 389/425 tons
Driver T.Tutt, Fireman R.Wilkes

Location	Mins Secs	Speed	(Schedule)	Engine working
Dover Hawkesbury St	00.00	(restart from Hawkesbury Box)		
Dover Priory	03.07		(4)	
Kearsney	09.28	30½	(1 in 132)	42% cut-off, full reg.
Shephersdwell	15.50	33½		
Adisham	20.43	72½		
Canterbury East	29.55	sigs stop	24)	
MP 60	33.16	34½		
MP 57	38.30	4½	(1 in 132)	37% cut-off, full reg.
Selling	41.06	55½		
Faversham	44.17	65	(38)	
Teynham	49.03	sigs 40*/ 54		
Sittingbourne	52.52	48½	(48)	
MP 43	55.10	37	(1 in 100/120)	
Rainham	60.09	54		
Chatham	65.39	slack	61)	
MP 32	69.13	39	(1 in 100)	
Sole Street	83.40	sigs 5*	(77)	
Farningham Road	93.08	71½		
Bickley Junction	103.08	sigs	(96)	
Herne Hill	124.23	sigs	(113)	
-		sigs stand		
Victoria	143.50		(120)	

30776 *Sir Galagars* on an up continental boat train that has been routed via Maidstone East, seen here after passing Shortlands Junction, c1958.
(Ken Wightman)

30795, newly overhauled and with a bogie tender, departs from Bromley South with a Ramsgate express, 25 May 1958. (R.C. Riley)

30794 *Sir Ector de Maris* surmounts the 1 in 100 from Beckenham Junction at Downs Bridge with a Victoria–Ramsgate express, 26 August 1957. (Ken Wightman)

30801 *Sir Meliot de Logres* takes a breather at the summit at Downs Bridge, near Shortlands Junction, with a down Ramsgate express, 9 August 1957. 30801 was one of the King Arthurs that kept its six-wheel tender to the end and never made it to the Western Section after the Kent Coast electrification, being withdrawn in April 1959. (Ken Wightman)

30804 *Sir Cador of Cornwall* departs from Bromley South with an express for Ramsgate, c1959. Although transferred to the Western Section, 30804 retained its six-wheel tender, being withdrawn in February 1962. (Ken Wightman)

30796 *Sir Dodinas le Savage* descends Sole Street Bank with a Ramsgate express, August 1958. This locomotive seems to have been particularly photogenic, appearing in many photos on all three sections of the Southern Region. Interestingly its name signifies it was an outsider or 'foreigner' to Arthur's court, 'le Savage' merely denoting that the knight lived in the forest rather than referring to his character. (Ken Wightman)

30776 *Sir Galagars* with an up freight and empty stock train approaching Folkestone Junction, 15 July 1954. (J.M. Bentley Collection/R.C. Riley)

30804 *Sir Cador of Cornwall* at Dover Marine with a boat train for Victoria, 2 August 1955.
(J.M.Bentley Collection/L.Hanson)

30798 *Sir Hectimere* with a Victoria-Folkestone train at Petts Wood Junction, on the last Saturday of steam working on the Kent Coast via Chatham, 13 June 1959.
(P.H.Groom)

30766 *Sir Geraint* on a night Kent Coast express, c1957. 30766 was the first King Arthur to be withdrawn in December 1958, after running 1,141,019 miles.
(Colin Garratt Collection/Rev A.W. Mace)

A six-wheeled tender Arthur on arrival at Victoria's No.2 platform with a train from the Kent Coast, c1957.
(Colin Garratt Collection/Rev A.W. Mace)

A six-wheel tender Arthur stands at Tonbridge awaiting departure with the Birkenhead–Deal cross-country train made up of Western Region rolling stock, as Schools 30919 *Harrow* approaches with a Folkestone–Charing Cross express, c1957. (Colin Garratt Collection/Rev A.W. Mace)

30805 *Sir Constantine* stands at Paddock Wood with a Tonbridge–Ashford stopping train, being overtaken by a Battle of Britain Pacific on a Dover express, while a class 'R' 0-4-4T stands on the left, c1955. (Colin Garratt Collection/Rev A.W. Mace)

30796 *Sir Dodinas le Savage* at Cannon Street with the 5.47pm to Dover, 5 June 1958. (J.M. Bentley Collection/J.H. Aston)

30796 *Sir Dodinas le Savage*, 'bulled up' for a Brighton–London railtour, runs down the Brighton route with a West Midlands–South Coast cross-country train of mixed stock, c1958.
(J.M. Bentley/P. Leavons)

An RCTS survey in August 1955 documented the activity of every mainline passenger steam locomotive on the Southern Region. The Eastern Section census was carried out on the 1 August and found the following turns worked by King Arthurs:

Stewarts Lane engines:
30764, 30773, 30774, 30791-30794 on Victoria–Kent Coast expresses
30767, 30769, 30770, 30772 on relief continental boat trains
30795 on a through train from the Kent Coast to Nottingham
30765 on 11.26am Victoria–Ramsgate and 3.48pm Deal–Charing Cross
30763, 30766 and 30771 in Eastleigh Works
Bricklayers Arms engines:

30799 working ferry vans
30800 and 30801 both on Charing Cross–Ramsgate trains
Hither Green:
30806 on the 8.50am Charing Cross–Ramsgate
Dover engines:
30775 and 30777 on relief continental boat trains
30776 and 30796 on Victoria–Kent Coast via Chatham trains
30798 on 8.54am Dover–Charing Cross and 1.57pm Victoria–Deal via Maidstone East
30797 on ECS Folkestone Junction–Victoria via Tonbridge

It will be noted from the above that more transfers had taken place with the Stewarts Lane stud of King Arthurs being increased to seventeen locomotives for the summer service. Whilst every King Arthur except for

the three in Works were noted active that day, this contrasts significantly with the Eastern Section West Countries and Battle of Britains, only twenty-three of the possible thirty-six being observed at work.

The survey of the South Western Section took place three weeks later on 22 August and found very similar results:

Nine Elms engines:
30455, 30456, 30780, 30781 on Bournemouth line expresses
30778 and 30779 on Waterloo-Salisbury semi-fast services
30457 on 8.22 Waterloo–Exeter and 1.55pm Exeter return
Salisbury engines:
30451 and 30453 on Exeter–Waterloo expresses
30450 and 30452 on Salisbury–Waterloo semi-fast services

30448 and 30454 at Eastleigh Works

30449 under repair on shed

Eastleigh engines:

30785 on a Bournemouth–Waterloo express

30784, 30788 and 30790 on Channel Island Boat Trains to Weymouth

30787 and 30789 on Portsmouth–Reading services

30786 under repair on shed

Bournemouth engines:

30783 on a Poole–Birmingham train as far as Oxford

30782 at Eastleigh Works

Of the twenty King Arthurs available, eighteen were observed in traffic that day compared with only

30790 *Sir Villiers*, with a purloined train of LM and Western Region stock, on a Waterloo–Bournemouth relief train, c1958. (MLS Collection)

30781 *Sir Aglovale* with an up freight from the Salisbury direction approaching Basingstoke, c1957. (MLS Collection)

30784 *Sir Nerovens* between Reading West and Basingstoke with a West Midlands– Southampton Docks freight which it will have taken over at Hinksey Yard (Oxford), c1958. (ColourRail)

forty-five of the fifty-eight West Countries.

As mentioned in the previous chapter about the N15s in BR traffic, in the summer of 1956 the King Arthurs were still very active in holiday relief trains, despite the fact that BR Standard 5s 73080-73089 had in the intervening year been allocated to Stewarts Lane and 73110-73119 to Nine Elms. The influx of the Standards on the Eastern Section had seen 30763, 30770 and 30771 sent to Basingstoke to replace withdrawn N15Xs, 30773 and 30774 allocated to Nine Elms and 30764, 30767 and 30791 transferred to Bournemouth. On 28 July, the Charterhouse Railway Society surveyed the South Western main line between Weybridge and Basingstoke from noon until about 7pm, and I have augmented their notes with data from a survey by Ben Brooksbank for the RCTS, who was at Basingstoke for part of the day before the school society

members appeared. On this summer Saturday, the Southern Region managed to run at least thirteen Ocean Liner Expresses – five in the down direction and eight on the up – both directions running in the peak traffic flows. Unusually, only one was hauled by a Lord Nelson. Nine were hauled by King Arthurs, including four by Urie engines. Three of the 70D N15s (30751, 30753 and 30771) worked boat trains in both directions.

In fact, King Arthurs of all descriptions were well represented during the day, with the following being noted by either the Charterhouse team or Ben Brooksbank - Urie Arthurs 30739, 30747, 30749, 30750, 30751, 30753; Eastleigh Arthurs 30448, 30449, 30454, 30456, 30457; and Scotch Arthurs 30763, 30765, 30771, 30773, 30774, 30779, 30780, 30781, 30782, 30788, 30789, 30790, 30791. Only Salisbury's 30449 and 30454 were on Waterloo–Salisbury semi-

fasts (I travelled from Woking to Farnborough and back behind each). The other twenty-two were all on Bournemouth or Exeter booked service or relief trains, or Ocean Liner Specials. However, two Urie N15s, 30751 on a down Ocean Liner Express and 30739 on a heavy Bournemouth–West Midlands service, were both struggling for steam and holding up a succession of trains behind them before arriving simultaneously at Basingstoke from opposite directions. Apart from those two, and a comment by the Fleet observer that 30774 on the 3.30pm to Bournemouth seemed to be riding like a 'bucking bronco', the general impression was that they were performing very competently. 30750 on an Ilfracombe-Waterloo eleven-coach train, was stopped outside Woking station awaiting a path, blowing off steam furiously and two of the meagre number of trains actually seen on time were both Arthur – hauled – 30763 on the 5.30pm Waterloo-Bournemouth and 30457 on the 7.09pm arrival from Bournemouth (for full details, see chapter 11).

The lack of flexibility of the King Arthurs with different types of tenders, particularly the restrictions imposed by the 3,500 gallon six-wheel variety, was ameliorated from 1955 when Urie 5,000 gallon bogie tenders started to become available from withdrawn engines, both N15s and N15Xs. The Drummond 'watercart' tenders were also needing heavy repair, so the Salisbury Eastleigh Arthurs got priority, 30448 receiving a tender from Urie S15 30503 in May 1955. 30457 was unique in getting a high-

The driver waits for the right away on 30454 *Queen Guinevere* as it stands ready at Salisbury to depart with a train for Exeter, September 1957. 30454 was provided with a bogie tender from Urie Arthur 30755 withdrawn earlier that year.
(John Hodge)

30451 *Sir Lamorak* departs Salisbury for Waterloo with an up relief West of England train, summer 1957. It received the bogie tender of N15X 32333 in January of that year.
(John Hodge)

30796 *Sir Dodinas le Savage* works the RCTS special from London to Brighton through Redhill, c1958.
(J.M. Bentley Collection)

sided 5,200 gallon tender from the recently withdrawn H15, 30490 in July. 30449, 30450, 30453 and 30455 received tenders from Urie N15s between November 1955 and December 1957, 30456 the last, in August 1958. However, 30451 had obtained the bogie tender from 32333 in February 1957 and 30452 from a Maunsell engine, 30785, in June 1957 (the latter was a swap from the condemned N15 30746 – not a withdrawal).

Once the six-wheeled tender Arthurs were made redundant on the South Eastern Section through electrification of the Chatham and Tonbridge routes, it became desirable to replace the tenders of those engines still deemed in good enough condition to transfer, to the South Western Section. Eight of those locomotives received four double bogie tenders off withdrawn Urie N15s, three from withdrawn Maunsell King Arthurs and one from the last Remembrance, 32331, between June 1958 and January 1961. In two cases it was the

Relocated from Hither Green and transferred to Salisbury in June 1959, 30796 *Sir Dodinas le Savage* has lost its sheen, and is entering its home location with a semi-fast train from Waterloo, July 1959. It received a final overhaul at Eastleigh in January 1960, when it received a bogie tender from a withdrawn locomotive and was itself withdrawn in March 1962.
(John Hodge)

30796 *Sir Dodinas le Savage*, now equipped with bogie tender, runs into Salisbury with a stopping train from the west, c1960.
(J.M. Bentley Collection/H.C. Doyle)

30452 *Sir Meliagrance* departing from Exeter Central with a stopping service for Salisbury, 29 June 1957. 30452 was withdrawn in August 1959 after running 1,494,011 miles. (R.C. Riley)

Salisbury Eastleigh
Arthur with bogie tender, 30449 *Sir Torre*, on the 2.54pm Waterloo-Basingstoke & Salisbury semi-fast, shuts off steam to brake for the Clapham Junction curve, 6 September 1958.
(R.C. Riley)

6.54pm Waterloo-Salisbury, first stop Woking, then all stations; the trains normally consisted of two three-coach sets plus the odd van or two. In 1957 and 1958, 30448–30454 were the engines used and, after the withdrawals of 30448, 30449, 30452 and 30454, three South Eastern Section engines, 30796, 30798 and 30799, were transferred to Salisbury to join the still active 30450, 30451 and 30453.

An Eastleigh engine would still come up with the afternoon van train, a service that a 71A King Arthur had commanded for over ten years. Occasionally one would deputise for a Lord Nelson on the heavy 6.4am Southampton Terminus–Waterloo commuter train, on which, frankly, it would outshine its theoretically more powerful peer. Otherwise, the Eastleigh-based engines would run local services to Southampton, Bournemouth and Weymouth, freight services from Southampton

second switch, 30798 and 30800 receiving tenders from 30450 and 30454, which themselves received the tenders from 30737 and 30755. 30794, 30797, 30799, 30801, 30804 and 30805 all kept their six-wheel tenders to the end, 30797 and 30801 being withdrawn when the Chatham route was electrified.

During their final couple of years on the South-Western Section,

the remaining King Arthurs were employed on Waterloo–Basingstoke–Salisbury semi-fasts, van trains and fast freights, with summer Saturday work in 1959 and 1960. Salisbury had two regular weekday turns to London, the 8.46am Salisbury arriving mid-morning and an early afternoon Salisbury semi-fast train, returning respectively on the 2.54pm and

30796 *Sir Dodinas le Savage* departs Seaton Junction with a stopping train to Exeter, with an M7 hovering off the Seaton branch train, 12 September 1959. (Ken Wightman)

Docks and Ocean Liner specials. The Bournemouth engines, 30764, 30771, 30772, 30782 and 30783, would work cross-country trains from Poole to Oxford and locals in the Bournemouth area.

The Nine Elms final allocation of 30763, 30774, 30778 and 30779 would go turn and turn about with a 70A Schools, after the Hastings dieselisation, on the 12.54pm Waterloo–Salisbury and were the booked power for the late night 11.15pm to Basingstoke, returning on a freight. 30455- 30457 were the regular steeds for the eleven-coach 6.09pm Waterloo–Basingstoke commuter train and the Basingstoke engines would alternate with

Standard 4 4-6-0s and Schools on the ten coach 5.09pm. After 1959, Basingstoke's 30765, 30773, 30777, 30793 and 30795 were the most frequent and effective performers, with 30777 *Sir Lamiel* dominating haulage of the train for months towards the end of 1959 and the first half of 1960. For a while, some of these had been based at Feltham which included the 5.09 diagram, but the turn and the engines moved to be a Basingstoke responsibility. In the summer of 1961 only 30798 was seen regularly on holiday relief trains – the 9.25am Wimbledon–Weymouth and 3.50pm return on at least five occasions. In their last summer, 1962, only 30770 appeared

regularly, on Bournemouth line relief trains, a couple of Ocean Liner Specials and a Channel Island Boat Train.

A colleague of mine and close friend, Colin Boocock, recounted to me some time ago his experience of a very high speed run with a King Arthur on a rolling stock test run and he has allowed me to recount part of it that is relevant here. I can do no better than use some of his own words:

'In summer 1960 I was working for the Southern Region CM&EE's Technical Development Office based in the carriage works at Eastleigh. We had been tasked with improving the ride of the

Salisbury's 30799 *Sir Ironside* slips under the Battledown flyover at Worting Junction with an afternoon Salisbury–Waterloo semi-fast service, the return working being the 6.54pm Waterloo–Salisbury, c1960. 30799 was transferred to Salisbury in June 1959 with 30796/8 to replace three of Salisbury's withdrawn Eastleigh Arthurs. (ColourRail)

A run-down Salisbury Eastleigh Arthur, 30452 *Sir Meliagrance*, at Salisbury station in 1957, having amassed 89,000 miles since its last heavy repair and a few days before entering Eastleigh for a 'light casual' repair, allowing it a further couple of years before withdrawal and replacement at Salisbury by King Arthurs from the Eastern Section. (John Hodge)

30448 *Sir Tristram*, on the 2.54pm Waterloo–Salisbury semi-fast train, strolls into Farnborough past the Railway Enthusiasts' Clubroom, July 1959. (Author)

Kent Coast electric buffet cars, and had No.S68009 as guinea-pig. The vehicle was marshalled in the middle of a BR Mark 1 four-coach locomotive-hauled set, modified with a through vacuum pipe so that braking could be done by the four Mark 1s and haulage could be by any available steam locomotive.

'The first run was from Eastleigh to Weymouth, outward via Bournemouth and return via Wimborne. Haulage was a BR Class 4 2-6-0 with 5ft 3in wheels. Maximum speed was 81mph down Hinton Admiral bank. The buffet car was instrumented with a home-made device using a timed, continuous paper roll on which were scribed ink traces that followed the lateral and vertical movements of the bogie spring plank relative to the bogie frame. Thus we could measure bounce, sway, pitch and yaw frequencies and magnitudes.

'The officers on this train were dissatisfied with the speed range achieved on this first run. A second run was organised, this time between Eastleigh and Basingstoke and return. Eastleigh shed provided 30804 *Sir Cador of Cornwall*, driven by one of the two Drivers Norton (not related) of Eastleigh shed, this one not being the known firebrand one.

'David Allcock, my immediate superior, asked the driver to achieve what high speeds he could because we wanted to study the coach's ride at all possible speeds. He must have sparked Driver Norton's imagination. As soon as we were on the main line, 30804 went like the wind, balancing out at 75mph most of the way up the 1 in 252 to Roundwood summit, and passing Worting Junction at an uncomfortable 83mph.

'At Basingstoke the engine was turned and the train made ready for the run back to Eastleigh. David Allcock and I were measuring train speed in parallel by counting the rail joints (me using C.J. Allen's declared method), timing the

30804 *Sir Cador of Cornwall* stands at Eastleigh station with a local stopping service, c1959. This engine was used on a special test run of Kent Coast electric stock, touching 91mph in the descent from Micheldever around this time, recorded by Colin Boocock who was then employed at the Works. (MLS Collection)

30804 *Sir Cador of Cornwall* on a down ballast train leaving Woking, June 1960. (Author)

passing of mileposts (David), and marking the times and distances on the moving paper sheet. Thus on this run we had three-way corroboration of the speeds achieved. I forget how fast we breasted the summit in Roundwood tunnel, but I do remember that we entered the Weston-Wallers Ash section at 90 and left it at 91mph, a speed maintained until just short of Winchester. We screamed through Winchester with the whistle at full blast, still going at 89mph.

'All through the speed range the carriage ride was not as appalling as in Kent. We judged this was because the track was in much better fettle than that on the North Kent main line. Adjustments were made to the bogie suspensions, and a second run was made the next day to the same timings with

the same engine and crew. The only difference was the presence of Stephen Townroe, District Motive Power Superintendent, Eastleigh, in the buffet car with the engineers. (Had word got around?) This time the uphill performance was again excellent, but the return run achieved "only" 83mph, just inside the SR's 85mph speed limit on main lines. Maybe Townroe's presence presaged caution! Two more runs on the same route used an S15 and a BR 4 4-6-0, but neither produced any exciting speeds.

'The project then switched from Eastleigh to Lancing Works. Several runs were made with different bogie configurations, under guidance from Jury Koffmann, a DM&EE bogie ride expert. Initially the next runs were again with steam haulage using the same

train formation, using Schools class 4-4-0s. All I remember of these was a maximum of 83mph somewhere on the up Brighton main line, and then passing through East Croydon's platform 1 loop at 60mph, the sideways lurch propelling me out of my toppling buffet car chair into the arms of Charles Shepherd, the Southern's C&W Engineer, whose normally robust response was muted to, "Lord, bless my soul!"'

The last King Arthur to receive a general repair at Eastleigh Works was 30451 in January 1961. The first King Arthurs, as opposed to the Urie N15s, to be condemned were 30454 and 30766 at the back end of 1958, the first having worked 1,421,676 miles and the latter 1,141,019. Then, when ten redundant Standard 5s were transferred to Nine Elms from Stewarts Lane in 1959, the writing was on the wall. Sixteen King Arthurs were withdrawn from service in 1959 and nine more in 1960, leaving just half the class

Six-wheel tender Arthur 30805 *Sir Constantine* leaves Weymouth with a train for Bournemouth, in the summer of 1959. It transferred from the Kent Coast in June 1959 and was withdrawn from Eastleigh depot in December 1959, before it could receive a bogie tender. (Ken Wightman)

– another twenty-seven – intact at the beginning of 1961. Fifteen more went in 1961 leaving the last dozen to survive to 1962. 30773, 30781, 30788, 30796 and 30804 went in the first half of the year and the final withdrawals were Salisbury's last pair, 30451 (the last Eastleigh Arthur) and 30798 in June, 30795 in July, 30793 in August, 30765 and 30782 in September and finally 30770 *Sir Prianius* of Eastleigh in November, having outlived its remaining Lord Nelson shedmates by a couple of months. The last rites were performed by 30782 *Sir Brian* on an enthusiast special in February, the LCGB *Kentish Venturer*.

The highest mileages were attributed to the ten Eastleigh Arthurs, with the highest of all, 1,606,428 being claimed by 30453

The former 'star' of the Salisbury–Exeter route in the 1930s and 'Pride of Stewarts Lane' in the 1950s, 30768 *Sir Balin*, was transferred to Eastleigh in June 1959 and worked from there until its withdrawal in November 1961. It is here seen on a relief Bournemouth–Waterloo express on a wet afternoon, in the summer of 1961. (ColourRail)

King Arthur himself. *Sir Lamorak* ran a creditable 1,579,556. I have seen a mileage of 2,010,095 claimed for 30453 and similar mileages for the 30448-30457 series, but these records will include the mileages run in Drummond G14 or P14 form. As only the tender can even be said to have been part of the rebuild, and all of those were replaced by Urie or Maunsell 5,000 gallon bogie tenders between 1955 and 1957, the crediting of these locomotives with the higher mileage is totally spurious.

The highest mileages by any Scotch Arthur were 30786's 1,389,62; 30789's 1,383,297; and 30784's 1,369,983. The former Brighton engines with their short runs, and later on the Eastern Section with maximum distances per trip of around eighty miles, struggled to reach the round million, with just seven of the fourteen achieving that milestone, with 30806 *Sir Galleron* highest at 1,127,096. The lowest mileage by any King Arthur was 30794 with just 903,663 miles, and the lowest Scotch Arthur was 30764 with 979,213 (the only one not to reach a million miles). As nearly all the

30770 *Sir Prianius* when it was the last surviving King Arthur, just two months before withdrawal and the class extinction on British Railways, 2 September 1962. (D. Clark)

King Arthurs were broken up at the Southern Region main works, only 30777 escaped to survive in preservation, being identified as one of the engines selected for the National Rail Museum. Initially the choice fell on 30453 itself, but it did not have its original Drummond 'watercart' tender and its frames were said to be badly fractured, so 777, the hero of the record *Atlantic Coast Express* – and in its last days, the known favourite of the Basingstoke drivers – was substituted. Its later career is outlined in chapter 13.

10.4: The N15Xs

At nationalisation, the seven N15X Remembrances were all based at Basingstoke from where they ran semi-fast and commuter services to Waterloo and Salisbury, and cross-country services to Portsmouth and Bournemouth from

32328 *Hackworth* in faded malachite green, waits for the road at Waterloo with the 5.9pm Waterloo–Basingstoke commuter train, c1949. (MLS Collection)

Oxford and Reading, sometimes exchanging with the GW engine at Basingstoke. On summer Saturdays they would augment the motive power available for the plethora of seasonal and relief trains, especially on the Bournemouth line.

As a schoolboy attending Surbiton County Grammar School between 1949 and 1951, if I got out of school early just before 4pm, I'd see one flash past the main up platform with a four-coach semi-fast from Basingstoke to Waterloo. I can still see 32330 *Cudworth* in my mind's eye, resplendent in malachite green, but with BR number on the cabside and 'British Railways' across the green tender. Then there were *Hackworth* and *Stephenson* in lined BR black livery and a vision one day of the 3.54pm Waterloo–Basingstoke pounding on the Down Through Line behind 32333 *Remembrance* itself, the only time I saw this engine, although nearly all the others were common – with the exception of 32327

32331 *Beattie* approaches Vauxhall with a Waterloo–Salisbury semi-fast train, 8 July 1950.
(Transport Treasury/R.C. Riley)

32327 *Trevithick* departs from Farnborough station with a semi-fast train from Waterloo to Basingstoke, 7 July 1950.
(Transport Treasury/R.C. Riley)

32329 *Stephenson* approaches the Farnborough stop on a down semi-fast Waterloo–Basingstoke train, probably the 5.9pm Waterloo commuter train, 7 July 1950. (Transport Treasury/R.C. Riley)

Trevithick which I never did see. In fact, all were repainted in the BR lined black 'mixed traffic' livery in the early 1950s.

Their reduced performance and power capability was recognised in that they were only classified 4P, whereas the N15s and King Arthurs were designated 5P (itself a somewhat pessimistic and conservative ranking given that the Lord Nelsons were granted 7P status). They were useful on the stopping and semi-fast services, though, as their initial acceleration was good, but they quickly became rough-riding as their mileage between heavy Works repairs

N15 30739 *King Leodegrance* is ready to depart from Oxford with a cross-country service for Bournemouth while N15X 32333 *Remembrance* awaits the next southbound cross-country train, 6 September 1952. (Transport Treasury/R.C. Riley)

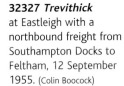

accrued. Given the right occasion and an enthusiastic crew they could put up a reasonable performance. On 23 June 1956, 32329 Stephenson, just before its withdrawal, was the motive power for a Stephenson Locomotive Society special train from London Bridge to Brighton. Despite a signal stop near Purley and checks at Haywards Heath and Preston Park, it got its seven-coach 228-ton train to Brighton in just over the round hour, with a net time of 55 minutes. Top speed was 76mph at Horley.

32327 *Trevithick* at Eastleigh with a northbound freight from Southampton Docks to Feltham, 12 September 1955. (Colin Boocock)

32329 *Stephenson* at Bournemouth awaiting departure with the 3.55pm Bournemouth–Waterloo, 7 August 1955. (Colin Boocock)

32333 *Remembrance* with a light parcels train at Eastleigh, c1953.
(MLS Collection)

32331 *Beattie* leaving Basingstoke with a cross-country train for Bournemouth, c1955.
(9 MLS Collection/N. Fields)

32328 was the first to be withdrawn in January 1955 and 32330 followed in August the same year. 32327 was involved in a collision with an electric train at Woking on 23 December 1955 while working the 7.54pm Waterloo–Basingstoke. It was badly damaged and was condemned the following month.

32332 was also withdrawn in January 1956, 32333 in April and 32329 in July, just one month after its special SLS tour. They had outlived some of the Urie N15s, which were then engaged on much the same type of work and were facing the influx of the ten BR Standard 5s, 73110-73119 and four Standard 4 4-6-0s, 75076-75079, which were allocated that year to the Southern Region, the Standard 5s being based at Nine

32329 *Stephenson* draws out empty stock ready to form a Basingstoke–Waterloo semi-fast train, 22 April 1956.
(Transport Treasury/R.C. Riley)

32333 *Remembrance* still in malachite green, on a freight at Bramshott between Reading and Basingstoke, 1950.
(ColourRail)

32329 *Stephenson* runs light from shed to pick up its working, a semi-fast train from Basingstoke to Waterloo, 22 April 1956. (R.C. Riley)

32329 *Stephenson* with a railway enthusiasts' special at Andover Junction, 1956. (R.C. Riley)

32331 *Beattie* on a special excursion to Windsor Riverside at Tulse Hill, 23 June 1957. (R.C. Riley)

The front end damage to 32327 sustained in its collision with an electric train at Woking whilst working the 7.54pm Waterloo–Basingstoke in 1955. (Colin Garratt Collection/Rev A.W.Mace)

Urie S15 30500 on a shorts goods train in the Surbiton area, c1949. (J.M. Bentley Collection)

Urie S15 30508 at a road crossing on Southampton Docks, c1950. (Colin Garratt Collection/Rev A.W. Mace)

Elms and the Standard 4s at Basingstoke.

The last survivor, 32331 *Beattie*, was withdrawn in July 1957 making the class extinct. The nameplate and memorial plaques from *Remembrance* were mounted and hung in Brighton Works.

10.5 The S15s

All of Urie's S15s entered the service of the nationalised British Railways in 1948, and were soon all painted goods black and renumbered 30496-30515. As a trial, Feltham's 30496-30498 were replaced by Q1 0-6-0s, 33009-33011, in April 1948, but were not a satisfactory substitute and the S15s had returned to Feltham from Nine Elms by December 1949. I assume that lack of

Urie S15 30503 departs Andover Junction with a Basingstoke–Salisbury stopping train, 7 September 1957. (Colin Garratt Collection/Rev A.W. Mace)

weight for braking purposes was as much an issue as power. Throughout the 1950s and up until 1963-4, they were the main Southern Region freight engines and worked steadily, if usually unsung, reliably and without fuss. They would work cross-London freights to Willesden, Neasden and Reading as well as over the Southern main lines. The entire class of twenty engines was based at Feltham motive power depot.

The pre-nationalisation distribution of the Maunsell S15s remained static until January 1950 when 30827 joined 30828-30832 at Salisbury and, in June 1951,

30505 enters Salisbury with a three-coach local stopping train from Yeovil Junction, July 1957. (John Hodge)

30510 heads a freight down the gradient from Litchfield summit to Winchester for the Bournemouth line, c1958. (MLS Collection)

30499 heads purposefully through the London suburbs with a freight for Feltham and the cross-London lines, c1959. (MLS Collection)

30510 enters Southampton Central with a local passenger service from Eastleigh, c1960.
(John Scott-Morgan Collection/A. Gosling)

30499 restarts a heavy freight on the slow line at Farnborough past track men, c1959.
(John Scott-Morgan Collection/A. Gosling)

30511, ex-works in April 1962, with an up goods passing Basingstoke.
(J.M. Bentley Collection)

The last survivor of the Urie S15s, 30512, taking water at Southampton Central, before moving off with a heavy freight for Feltham Yard, c1963.
(John Scott-Morgan Collection/A. Gosling)

30507 battles against a strong south-west gale as it approaches Basingstoke with a down goods, c1959. (Ken Wightman)

30496, the last of the Urie S15s to be built, passes through Eastleigh northbound with a freight from Southampton Docks, 6 July 1955. (Colin Boocock)

A busy scene at Eastleigh as 30500 passes through with a freight for Southampton Docks, with an M7 in the up platform and a T9 4-4-0 over the far side, 6 July 1955.
(Colin Boocock)

30502 passing Allbrook on an up goods for Feltham, 17 March 1962. This was one of the first examples of the Urie S15s to be condemned, in November 1962, having run 1,276,916 miles. (Colin Boocock)

30507 approaches Basingstoke with a down heavy freight for the Salisbury line, c1959. (Ken Wightman)

Ex-works 30501 emerges from the Clapham Junction curve with the 3.54pm Waterloo–Basingstoke semi-fast train, on Saturday, 20 June 1959. (R.C. Riley)

30511 stands at Southampton Terminus ready for departure, June 1963, just one month before withdrawal. (Ken Wightman)

30823-30825 and 30847 also left Exmouth Junction for Salisbury. Three of the Feltham engines, 30835-30837, left Feltham for the Central Section at Redhill to work ballast trains between Reading and Ashford. The Maunsell engines, like the Urie examples, also worked passenger trains at peak periods including Ocean Liner specials. A couple of runs in the early 1950s with Maunsell S15s on heavy passenger trains were timed and the records still exist. 30836, then still at Feltham before its transfer to Redhill, worked a ten-coach plus vans relief boat train weighing 395 tons gross to Southampton Docks. It started slowly, taking twenty-four minutes to pass Surbiton, but covered the following thirty-six miles to Basingstoke at an average speed of 62mph, with a top speed

30839 with an up local service between Gasworks Junction and Bournemouth Central, 22 August 1953. (Colin Boocock)

of 67mph through Hook. Speed in the mid-60s down the bank to Eastleigh nearly held schedule, and, with a final signal check at Northam, the arrival at the Docks was just six minutes late. 30841, also of Feltham, found duty on a Saturdays 8.35am Waterloo–West of England as far as Salisbury with a twelve-coach 435 tons gross train. The start was even slower, taking a full ten minutes to reach Clapham Junction, but after a Surbiton stop speed had reached the sixties by Weybridge, and again after a p-way slowing at Brookwood 68mph was attained through Fleet. The engine was now warmed up and ran the thirty-six miles from Basingstoke to Salisbury in even time with a top speed of 71mph through Andover,

30839, with its original eight-wheel tender, on a down local stopping train at Eastleigh, 1954. (Colin Boocock)

A Salisbury Maunsell S15, 30824, stops at Chard Junction with a Salisbury–Exeter local train, c1960. (MLS Collection)

30830 waits for the road at Salisbury with a down freight, 1957. (John Hodge)

the highest recorded speed behind an S15 at that stage. I am surprised at the slow start from Waterloo of both these trains – I would have thought with their power and smaller wheels, initial acceleration would have been one of their strong points on runs like these.

On Monday 24 June 1957, 30505 was recorded in charge of the 6.35pm Exeter–Waterloo, presumably from Salisbury only. They sometimes worked cross-country trains to and from the Southern Region via Clapham Junction to Acton, Willesden or Brent.

They would find themselves on passenger work at summer weekends – in force on 19 July 1958 when 30503 worked the 11.22am Waterloo-Bournemouth, and 30513 on the 11.35am, although both lost twenty minutes in running. 30499 and 30512 were on afternoon services from Bournemouth to Waterloo and in the following month 30508 worked an excursion to Bognor, retiring to Brighton shed for servicing before returning. There was a Wimbledon starter for the Bournemouth line that was often turned out with a Feltham S15 and on summer Saturdays into the early 1960s, the lunchtime Waterloo-Salisbury semi-fast – a Nine Elms King Arthur or Schools during the week – would be a Urie S15. They also worked a few stopping trains in the Bournemouth–Weymouth area and sometimes came up from Eastleigh on a late evening Saturday slow train to Woking before the last dash for Waterloo. After the last of the H15s had gone by the end of 1961, they could occasionally be found on any of the Waterloo–

30830 with a down freight between Salisbury and Exeter, c1960. (MLS Collection)

30824 with a westbound perishable freight from Southampton, passes Totton, c1962. (MLS Collection)

Basingstoke semi-fasts, which by then were normally hauled by a King Arthur, Schools, Standard 4 or 5 4-6-0. The Feltham Maunsell engines in particular, after the mass withdrawal of Lord Nelsons, King Arthurs and Schools classes in 1962, frequently found themselves on heavy commuter trains to Basingstoke or Salisbury and the lighter semi-fast services if Nine Elms' Standard 5s were in short supply.

A number of the Urie engines obtained tenders from other withdrawn locomotives in the mid-1950s. Urie high-sided 5,200 gallon tenders from the H15s were assigned to 30498, 30503, 30504, and 30507 and Urie 5,000 gallon tenders went to 30505, 30506, and 30508-30510. Some of the Maunsell engines also changed tenders – 30847 received an Ashford six-wheel 3,500 gallon tender when

Salisbury's 30829 stands at Basingstoke with an afternoon Salisbury–Waterloo semi-fast train, c1957. The S15 will return on the 6.54pm Waterloo–Salisbury, a diagram the 72B S15s worked regularly until the stud of King Arthurs at the depot was strengthened after the Kent Coast electrification in June 1959. (MLS Collection)

30825 brings a Salisbury–Waterloo semi-fast service under Battledown flyover at Worting Junction, c1957. (J.M. Bentley Collection)

transferred to Redhill in 1960 and 30833 got a Schools 4,000 gallon tender in May 1962 from 30908 after it was withdrawn the previous year. 30837 also received a Schools tender from 30912 when that engine received a tender from a withdrawn Lord Nelson. As late as December 1963, 30825 was coupled to a flat-sided 5,000 gallon tender, as built for the 1936 series.

The allocation of the Maunsell S15s in the 1960s was:

Feltham: 30833, 30834, 30838-30840

Salisbury: 30823-30832, 30847*

Exmouth Junction: 30841-30846

Redhill: 30835-30837, 30847*

(* 30847 transferred to Redhill from Salisbury in 1960)

The S15s were only displaced from their freight duties by the incursion of the Crompton Type 3 D6500 diesels, which arrived at Eastleigh and Feltham in force in the latter half of 1962 for crew training. Freight services on the South-Western Section were reorganised for diesel haulage from 4 February 1963. Only the freight services west of Salisbury remained steam – that route being transferred to the Western Region with the Exmouth Junction fleet of S15s and Bulleid Pacifics. The remaining S15s there were replaced by 'N' Moguls and 30841-30844 were returned to work from Feltham after a short stay at Salisbury.

The first of the Urie S15s to be withdrawn were 30502, 30504 and 30505 in November 1962. Other withdrawals came slowly however, with the next, 30513, in March 1963. 30496, 30498, 30499 30501, 30506-30508 and 30512 were noted on passenger trains during that summer, though just three were left

The last survivor, 30837, with ex-Schools tender, on railtour duty, with the Locomotive Club of Great Britain (LCGB), 1965. 30837 was withdrawn in September 1965 having a final mileage of 911,016.
(John Scott-Morgan Collection/J.M.Bentley)

at the end of the year. In December 1963, these three, 30499, 30506 and 30512, were all engaged on Christmas Post Office and parcels trains, although the first two were condemned immediately afterwards. The last to be withdrawn was 30512 in March 1964, after it had performed enthusiasts' specials, including a Mid-Hants *Hayling Farewell* Rail Tour which it powered from Waterloo to Fratton on 3 November 1963.

The first withdrawal of a Maunsell S15 took place in November 1962 at the same time as its older Urie sisters – 30826 of Salisbury. Presumably an edict had gone out with the arrival of the diesels for freight work that no further S15s were to receive heavy overhauls. 30846 was condemned in January 1963 and 30829, 30831 and 30845 went later in the year.

Fourteen Maunsell S15s were withdrawn in 1964 and the last seven were condemned in 1965, the last three, 30837, 30838 and 30839, being withdrawn simultaneously on 19th September. 30837 worked a number of enthusiast specials before its official withdrawal.

Despite nearly all their work being on freight, and being only five years older than the King Arthurs, they put up very similar mileages to the Scotch Arthurs over their forty-two year life. The final survivor, 30512, attained 1,291,002 miles and the preserved 30506 1,227,897. The last to be built, 30496, reached 1,277,029 before its withdrawal in June 1963. Most of the rest were around the 1.1–1.2 million mile mark. 30497, 30507, 30509 and 30514 were sold to George Cohen's scrapyard at Kettering. The last three condemned

The first Maunsell S15 built in 1927, 30823, on a down Salisbury–Exeter stopping train at Seaton Junction, 12 September 1959. A former GW 64XX 0-6-0PT has replaced the M7s on the Seaton branch.
(Ken Wightman)

Despite their shorter life, the Maunsell S15s built in 1927 attained similar or even higher mileages than the Urie engines, with 30823 the highest at 1,411,643 miles. 30824-30827 and 30831 all exceeded 1.3 million miles and 30828-30830 and 30832 exceeded 1.2 million. The 1927-built locomotives for the Central Section ran between 800,000 and a million miles and the 1936 engines ranged from 931,829 (30847) to 781,397 (30840). Only seven of the Maunsell engines were broken up at Eastleigh, the remainder being sold to scrapyards – five to Cashmore's, one to J.Buttigieg, two to Bird's at Risca, two to Bird's at Morriston and two to Shipbreakers at Queensborough. 30825, 30828, 30830, 30841, 30844 and 30847 were sold to Woodham Brothers at Barry and thus five of them, 30825, 30828, 30830, 30841 and 30847, were bought for preservation – only 30844 missed out (see chapter 13).

were sent to Woodham Brothers in Barry, as Eastleigh had their hands full breaking up many of the passenger engines. 30512 was cut up, but 30499 and 30506 were rescued and were preserved, which happened to none of the Eastleigh-stored engines.

30824 heads a down freight at Broad Clyst, nearing Exmouth Junction, 6 July 1961. (R.C. Riley)

1936-built Maunsell S15 30841, with eight wheel flush-sided tender, (later to be preserved as 841 *Greene King*) double-heading 'U' mogul (rebuilt 'River') 31791 at Seaton Junction on a Salisbury–Exeter stopping train, 12 May 1959. (Ken Wightman)

30839 on a passenger train near Deepleap Wood between Gomshall and Dorking on the Guildford–Redhill line, 9 April 1964. (David Clark)

Chapter 11

PERSONAL REMINISCENCES

I lived on the Southern Railway and BR's Southern Region for twenty-six years from my birth in May 1938 until my first permanent railway employment as a stationmaster at Aberbeeg in the Western Valley above Newport in May 1964. Despite no obvious connection to railways in my family, I was fascinated by steam trains from a very early age and apparently used to bully my mother into walking me in my pushchair to Esher Common to watch 'proper' trains when I was less than three years old. My father travelled a lot by train during the war and my mother would take me and my younger sister to see him off on his various travels in uniform.

T14 460 passing Surbiton with a Waterloo – Plymouth express in the 1930s at the point the author regularly 'spotted' this and other T14s on the 3.54pm Waterloo–Basingstoke in 1949/50. (Mike Bentley)

The first (and only) one I can remember must have been about 1943 when we saw him off to his camp on Salisbury Plain. My dad always took me to see the engine (I suppose this is the connection that aroused my interest) and on that occasion the train was hauled by a Salisbury King Arthur, 454 *Queen Guinevere* – so my father told me in later years – about the only engine name he ever remembered. I can see the malachite green engine in my mind's eye still and remember the awful rumpus I kicked up when we got back into the train to bid him farewell. I thought we were going to get carried off with him, and I screamed until my poor mother had to lift me off onto the platform well before the train's departure time.

Unfortunately, holiday travels in the immediate post-war period were by coach – we couldn't afford train fares until a West Country journey in 1952 – apart from a short trip on the electric *Brighton Belle* in 1946. I'd been encouraged to start train-spotting on that brief Brighton holiday, but can't remember seeing and noting numbers of any of the Southern 4-6-0s until

I started secondary education at Surbiton County Grammar School in September 1949. We were then living at East Molesey and my daily trips to school started by local electric from Hampton Court. I used to join a group of other train-spotters outside the school before classes started, but then the only engines that passed were Bulleid Pacifics or Lord Nelsons. After school, however, it was a different matter. I and a group of friends would get back to Surbiton station between five and ten past four, and after seeing the Up ACE with its Merchant Navy, would watch with anticipation the 3.54pm Waterloo–Basingstoke burst from under the high bridge under the Ewell Road and pound past on the middle road. As often as not during my first year there, the motive power would be a

filthy black lumbering Drummond T14, but as the number of these dwindled, H15s, Urie N15s and King Arthurs would make more frequent appearances. Looking back, I think it must have been a Nine Elms turn, which would have supplied any of the 30482-30491 series (my 1949 printed Ian Allan ABC shows that I'd 'copped' all those ten engines). Urie N15s were frequent and I remember distinctly *Joyous Gard, Maid of Astolat, Tintagel, Etarre, Linette* and *Melisande*. King Arthurs appeared too, most still in their malachite green livery but with British Railways numbers and lettering – *Sir Galagars, Sir Lionel, Sir Menadeuke* and *Sir Uwaine* are the most memorable. On one occasion – and only one – we were overjoyed to see 32333 *Remembrance* itself on that train. We saw all the other

N15Xs, but only if we left school early and caught sight of a semi-fast from Basingstoke dashing through the Up Main Line just before four o'clock which invariably had a Basingstoke N15X – 2330 *Cudworth* in malachite green, and still with SR numbering and lettering, was the most common.

Only a few moments later, the up Eastleigh van train would appear, accelerating hard after it had slowed to regain the main line at Hampton Court Junction and follow the ACE. This was clearly booked for an Eastleigh depot H15 (30473-30478) or King Arthur, and my ABC soon got peppered with underlinings of the King Arthurs, for it would often get ex-works engines or engines from other sheds that Eastleigh had on hand, especially London area engines to

30806 *Sir Galleron* on the Eastleigh–Nine Elms van train between Brookwood and Woking ten years later than when the author regularly saw this train with similar power after leaving school at Surbiton, June 1960. (Author)

to join my uncle, aunt and adult daughter on a visit to their cousins in Axminster. Unfortunately we went by car – my uncle had a timber business and owned a Ford V8 into which we all piled. I was disappointed – I think they thought a car ride would be a treat for me. However, we made a day trip to the seaside at Exmouth, and as – with their cousins – we were too many to get into the car, some of us had to go by train. I volunteered! I and the daughter of the Axminster household (too old for me!) picked up a Salisbury–Exeter local with a Salisbury S15, 30826, and we went from Axminster to Sidmouth Junction, where we alighted for a run via Budleigh Salterton to our destination behind an M7. After a day paddling and eating candyfloss (my first experience of that), we arrived back at Exmouth only to find we had just missed the train to Sidmouth Junction. I promptly

S15 30826 at Yeovil Junction with a Salisbury – Exeter stopping train, c1961. 30826 was the first Maunsell S15 to be withdrawn in December 1962, although it had run 1,364,577 miles in traffic. The author travelled behind this engine on a similar train some ten years previously.
(Mike Bentley Collection)

be worked home. As far as I can remember though, it was always a 2-cylinder 4-6-0 – only twice was it not, and I remember those clearly for they were rarities; a Drummond L12 4-4-0 30424 and an Eastern Section Schools, 30917 *Ardingly*, the latter obviously after Works attention.

Around 1950, I was invited

30787 *Sir Menadeuke* which hauled the author's train from Surbiton to Salisbury in August 1952, seen here at Eastleigh five years later, 7 September 1957. 30787 was one of the first King Arthurs to be withdrawn in February 1959, having run 1.3 million miles.
(John Hodge)

persuaded my companion to join the M7 (30023) on a train for Exeter. I had no idea how long we would have to wait there – I was happy train-spotting while my companion adjourned to the refreshment room – but a stopping train for Salisbury eventually turned up hauled by that city's H15, 30330. I think I was a bit mystified – the only H15s I'd seen before were the Nine Elms ones – but I was happy enough to have got another 'cop' and we duly arrived back at Axminster to be met by the relieved remainder of the family party.

My next direct experience with a Southern 4-6-0 was on our outward journey to our fortnight's holiday destination of Paignton in August 1952. We travelled with all our luggage from Hampton Court to Surbiton to catch one of the early morning relief trains that stopped at Surbiton to pick up passengers from London's southern suburbs. A succession of trains called between 7.30 and 9am, and ours turned up behind a Nine Elms King Arthur, 30787 *Sir Menadeuke*. I was mildly disappointed as I'd hoped for a Bulleid pacific and it wasn't even a 'cop', but I was looking forward to Exeter and my first travels behind Great Western engines. The Arthur went well enough – it was before I attempted train timing – and I was pleased when we exchanged 30787 for Salisbury's blue Merchant Navy 35007 *Aberdeen Commonwealth* at the halfway point. The journey back after my paradise holiday (our local beach at Goodrington backed on to the Paignton–Kingswear line) started from Exeter St David's behind a Battle of Britain, but we had to change at Basingstoke as this

service did not stop at Surbiton and the last leg of our holiday journey was behind another of Salisbury's H15s, 30332. The following year, my family accompanied a group of Girls' Life Brigade girls for their annual camp on the floor of Newton Abbot Methodist Church (my aunt was Captain) and we had a party reservation from Surbiton to Exeter St David's. By this time I'd had a surfeit of Bulleid Pacifics with few left to cop, and I'd grown fonder of the King Arthurs. The same train we had caught the previous year came in with 30744 *Maid of Astolat* and I was very disappointed that we had to let it go, as our reservations were on the next train. I was more disappointed when it turned up with blue Merchant Navy, 35015 *Rotterdam Lloyd*, and even more disappointed when we failed to change engines at Salisbury, with 35015 just taking water there and working throughout. How fickle are the likes and dislikes of the young! Now I'd give anything for a run with an old Urie N15 and a blue unrebuilt Merchant Navy – equally. But on the way home, after a second week in our B&B in Paignton, 34032 *Camelford* was abandoned at Basingstoke for yet another Salisbury H15, 30334, on the local to Surbiton.

In 1954 I was entrusted with chaperoning a nine-year-old girl, who'd been staying with my aunt over the Easter holiday, back to Salisbury where we were to be met by the girl's father. We joined the 9.30am from Waterloo at Surbiton where it arrived behind Nine Elms' 30858 *Lord Duncan* which meandered painfully slowly to

Basingstoke, where we found 30781 *Sir Aglovale* waiting in the bay platform with a local for Salisbury. I brightened up at this (*Sir Aglovale* was a cop – one of my last King Arthurs) and it went much more energetically – it was only three coaches though. Having delivered my charge safely and got a copy of *Railway Magazine* given me for my trouble (hardly!), I found that *Sir Aglovale* was to be my steed home after spending the rest of the day spotting on the station. It was a better load though and the semi-fast train made a respectable speed and noise on the way home, while I devoured the magazine articles.

That summer, the family holidayed in Sidmouth and the fortnight is mainly memorable for runs behind many M7s in the area, and a few of the new BR Standard 3 2-6-2Ts, the first of which (82010-82019) had just been allocated to Exmouth Junction. On our return home, we travelled to the junction behind the Sidmouth regular branch engine, 30025, and joined with an Exmouth/Budleigh Salterton portion to be hauled from Sidmouth Junction by 34026 *Yes Tor*. To my surprise, we changed engines at Salisbury and I was astonished and disappointed that our engine for the non-stop run to London was only a 1924-built H15, 30523. The run was dire too. We were clearly short of steam for most of the way and I remember the run mainly for the filthy pall of brown smoke that kept enveloping our compartment as the H15 waffled along at a steady 40-50mph. We were very late arriving at Waterloo and I remember we'd missed our planned connection home.

7.38am Waterloo–Ilfracombe SO, August 1956
Salisbury–Exeter Central
30449 *Sir Torre*
11 coaches, 355/385 tons

Location	Mins Secs	Speed	Schedule
Salisbury	00.00		3L
Wilton	07.30	35	
Dinton	14.51	60	
Tisbury	19.11	54	
Semley	24.47	53	
Gillingham	29.42	77 / pws 20*	
Templecombe	38.32	50	
Milborne Port	42.05	35	
Sherborne	45.39	80	
Yeovil Junction	49.34	69/76	
Sutton Bingham	52.46	40	
Crewkerne	59.27	70	
Hewish	-	58/52	
Chard Junction	67.08	70	
Axminster	71.31	76	
Seaton Junction	74.19	67	
MP 150	-	43	
MP 152½ (Tunnel)	83.11	26	
MP 153½	-	28	
Honiton	86.34	51	
Sidmouth Junction	91.24	62	
Whimple	94.54	69	
Broad Clyst	97.59	75	
Pinhoe	99.41	56	
Exmouth Junction	-	½ minute signal stand	
Exeter Central	106.00	(98 mins net)	3L

Eastleigh Arthur 30449 *Sir Torre* on a Waterloo–Salisbury train passing Surbiton c1952, about the time it 'rescued' the Waterloo–Ilfracombe Summer Saturday express from the failed West Country. (Robin Russell)

I'd attempted to time part of that run (my new schoolfriends I'll tell you about in a minute had taught me to time trains using the second hand of my watch and the counting of rail joints) but as just described, 30523's exertions were not worth the effort. A couple of years later we went to Ilfracombe for our family fortnight, and our engine from Surbiton where we joined it was 34009 *Lyme Regis*, planned to work through to Exeter Central. However, it became obvious as we approached Salisbury that all was not well at the front end (I learned at Salisbury that its injectors had failed), and I got out to see 34009 disappearing instead of taking water. A long wait climaxed in the 'watercart' tender of 30449 *Sir Torre* backing from the shed onto our train, sulphuric yellowy-green smoke pouring from the chimney as the fireman attempted to break up the fire and prepare for the difficult terrain ahead. Clearly 30449 had been standing unattended on shed pilot or standby duties. I decided to try out my newly acquired timing skill and noted with some disappointment that our start to Wilton was dreadfully slow and a repeat of the 30523 run two years earlier was in my mind. How wrong I was! Once the fire had been sorted out, the engine livened up and the noise from the front-end reverberated down the train as we accelerated the eleven-coach express to Semley summit in a quite remarkable recovery. I print my log of a 1956 run over the Salisbury–Exeter main line with an Eastleigh Arthur in full, a rare opportunity to replicate the running of these engines in their heyday before the Second World War.

This run was totally unexpected and was up to pre-war standards, when King Arthurs monopolised this road with similar loads. After the slow start, the climb to Semley was exceptional as were the climbs to Hewish which was rushed and to Honiton itself. But for the signal check approaching Exeter Central and the crawl into the station, 30449 and its game driver and fireman would have recovered the time lost in changing engines.

In 1951 I left the Grammar School and received one of the three annual Surrey County Council assisted places at Charterhouse Public School. Despite being told that trainspotting was not an acceptable pastime for posh public schoolboys, in fact there were a group of half a dozen fellow enthusiasts who organised film shows, helped run a gauge 1 model railway and made visits to Longmoor Military Railway, Hayling Island or London sheds. We certainly visited Old Oak Common, Stratford and Camden, but the Southern Region interest was the trip to London for we invariably changed from our electric at Woking to sample main line steam power. Our usual train to Waterloo was the 10.30am Exeter which would have an Exmouth Junction Merchant Navy that would go back on the 5 o'clock. However, on one occasion, around 1954, a Basingstoke–Waterloo semi-fast was more convenient and I was delighted when 30755 *The Red Knight* hustled in with its four Maunsell coaches. The departure, I remember, was electric too, the

acceleration of this light load to Byfleet quite extraordinary – getting the same feeling as you do on an aircraft take-off. We stayed on the slow line, but our progress was hectic. I'm afraid I failed to time the train, but speed through Hersham was certainly in the upper 70s if not a full 80, although we had to brake at Esher as we went main line from Hampton Court Junction. The result of this haste was a long signal check at the throat of Waterloo station waiting for 30449 on the 2.54pm Waterloo–Salisbury to vacate our platform.

Fellow club members Philip Balkwill and Conrad Natzio devised an ambitious survey to take place on 28 July 1956, 'Black Saturday' as it was known, at the end of the summer term. They

were both avid train timers and got a group of us dispersed between Weybridge and Basingstoke with synchronised stop-watches to note passing times at five key locations between 12 noon and around 6.30pm. I've referred to this survey in previous chapters, noting the availability of the Urie and Maunsell King Arthurs compared with the Bulleid Pacifics and Lord Nelsons, but I thought it would be interesting to extract the Southern 2-cylinder 4-6-0s' data from the rest as relevant to this book

Only the pair of observers at Woking were present the whole time, but although the above is sketchy, it gives a rough 'feel' of a typical Summer Saturday in the mid-1950s. The times are accurate to the second, but the speeds are

30779 *Sir Colgrevance* passing Woking on a summer Saturday relief train to Bournemouth, July 1959. 30779 was withdrawn at the end of this month despite it being outwardly in excellent condition. (Author)

Down Services

Train	Loco	Load	Weybridge	Woking	MP30. ¼	Fleet	Basingstoke	Remarks
Dep time (pm) from Waterloo								
Ocean Liner Exp	30791 71A	13 chs	2/17 44 70	2/22 26 60	2/30 25 45	2/36 55 65	2/49 25 60	N/A - working hard
1.24 to Salisbury	30497 70B	6 chs	-	2.44 00 (dep)	-	-	-	49L (Slow line)
2.54 to Salisbury	30454 72B	6 chs		3.41 00 (dep)-	60	-	-	14L (Slow Line)
3.20 to B'mouth	30476 71A	12 chs	4/00 47 55	4/07 51 50	4/15 07 45	4/24 10 50	4/36 08 55	18L short of steam
ECS	30456 70A	13 chs	4/17 44 60	4/24 25 20*	4/35 55 40	4/44 01 50	5/16 56 0*	N/A *outside B'stoke
3.30 to B'mouth	30774 70A	11 chs	4/20 57 65	4.26 40//4.28 48	4/38 25 50	4/45 01 65	4.57 32	25L loco riding wildly
3.54 to B'stoke	30749 70D	8 chs + 3 vans		4.44//4.46	-	-	-	19L 0* Woking (ECS)
Nine Elms freight	30491 70A	40 wagons		5/58 20	-	-	-	To Soton Docks
5.30 to B'mouth	30763 71A	12 chs	5/53 41 70	5/58 52	60	6/05 22 50	6/12 10 70	1E
ECS	30840 70B	11 chs		6/10 40	-	-	-	N/A (Fast Line)

Up Services

Train	Loco	Load	Basingstoke	Fleet	MP30. ¼	Woking	Weybridge	Remarks
Arrival time (pm) at Waterloo								
1.42 ex B'mouth	30782 71B	12 chs	-	1/29 41 60	-	2.07 0*/30*	-	55L *before Woking
2.37 ex Ilfracombe	30750 70A	11 chs	-	2/05 18 65	-	2/26 18 10*	2/35 26 30*	18L * before Woking
Ocean Liner SPL5	30751 70D	8 chs	3/47 12 65	3/57 45 70	4/03 12 65	4/08 12 80	4/12 37 75	Whistle shrieking (Wok)
5.26 ex Salisbury	30449 72B	6 chs	4.15 32	4.33 15//4.34 03	4/46 40 50	5.02 42//5.03 39	5/16 24 10*	2 E app Wok, 7L (SL)
Ocean Liner 302	30753 70D	11 chs	-	-	-	5/22 55	-	N/A - Pullman car incl
5.53 ex B'mouth	30780 71B	10 chs	-	-	-	5/34 40*	-	11L
6.03 ex Lymington	30747 71A	10 chs	-	-	-	5/44 0*/25*	-	11L
Ocean Liner 305	30771 70D	12 chs	-	-	-	5/56 20*	-	N/A
6.55 ex Salisbury	30512 70B	6 chs	-	-	-	6.31//6.33	-	7L (Slow Line)
Ocean Liner 311	30788 71A	8 chs	-	-	-	6/38 30*	-	N/A Loco ex-works
7.09 ex B'mouth	30457 70A	10 chs	-	-	-	6/40 35*	-	1L

just estimates assessed by the observers. Few of the trains got through Woking unchecked. The most interesting performances above were that 30763 on the down and 30457, on the up were virtually the only punctual trains all day, 30791 had a thirteen-coach Ocean Liner Special, 30751 hurtled through Woking at an estimated 80mph with an up boat train, by far the fastest of the day at that point, 30747, was on a Lymington Pier train normally restricted to 4-4-0s or 'U' moguls and 30476 was making a dreadful hash of the Merchant Navy rostered 3.20pm Waterloo–Bournemouth. RCTS observer Ben Brooksbank was at Basingstoke and added some additional sightings in the morning

– 30751 *Etarre* on a down Ocean Liner Express took ninety-seven minutes to get to Basingstoke, clearly in trouble for steam, with the four portions of the down *Atlantic Coast Express* bunched up behind it, and 30739 *King Leodegrance* staggered uphill to Litchfield summit with a thirteen-coach Bournemouth–Oxford-Birkenhead train, running fifty minutes late and holding everything up behind it. 30751 had obviously got its act together for the return Ocean Liner express. And the last N15X, 32331 *Beattie* took over a Birmingham–Bournemouth relief at Basingstoke from a Western engine.

I was given charge of a master's ancient 8mm cine camera and left my comrade alone at Woking while I

travelled on the 2.54pm Waterloo to Farnborough behind 30454, filming passing up services. I returned almost immediately behind another Eastleigh Arthur, 30449, on an up semi-fast from Salisbury. Regrettably, I have no idea what happened to the rolls of film I took.

I left Charterhouse in December 1956 and had a nine-month gap before enrolling as a student at University College London. I obtained temporary employment as a relief clerk at Old Oak Common motive power depot and between January and August 1957 travelled daily from Hampton Court to Willesden Junction. There was, conveniently, an early morning steam train, the 6.39 Basingstoke, which stopped at Surbiton, and I made it my frequent practice to alight from my suburban EMU there and transfer to the steam train for its non-stop run to Waterloo. This train was always hauled by one of the massive Urie H15s that were based at Nine Elms and during those months I amassed over fifty runs behind the class, involving all eight of the locomotives that were still extant – 30485 and 30490 had been scrapped in 1955. I'd had just one run behind each of those H15s during days in which I'd travelled up to London trainspotting with a friend while I was still at the grammar school. I had ten runs each behind 30484, 30489 and the tapered boiler 30491, and managed three runs behind 30483 that was withdrawn in July that year, most of the others lasting until 1959 and two until 1961.

I can't say that the performance was anything to write home about. Although it was normally only a seven or eight-coach train, acceleration would hardly reach 60mph before the line speed limit of 60 (imposed for restricted signal distances) was reached at New Malden. It was at least satisfying to sample steam in one direction, though I later discovered I could return in the evening from Willesden Junction to Euston in the summer timetable behind one of the few remaining Rugby Compound 4-4-0s. There was just one occasion on which something other than a 70A H15 appeared. One morning I was pleasantly surprised to find a gleaming brunswick green N15, 30748 *Vivien*, at the front end. I can't remember if the run was any livelier than usual, but it was only my second run behind a Urie N15 and was to remain with 30755 the only

ones I sampled. I started college in September 1957 and travelled for a couple of months up from Hampton Court to Waterloo, but most of my lectures did not start until 10am and there were no stopping steam trains I could catch at that hour from Surbiton. However, on a Thursday morning I had a 9am start and decided to go just a little early and catch the 6.39 Basingstoke as I had during my Old Oak employment. Imagine my delight when I saw 30738 *King Pellinore*, then the last survivor of the class, approaching slowly, when suddenly the colour light signal on the Up Main changed from red to amber and the driver opened the regulator and left me standing open-mouthed on the platform. The train planners had cut out the Surbiton stop in the 1957-8 winter timetable!

My family moved to Woking in November 1957 and I exchanged my Hampton Court–Warren Street season ticket, paid for from my Surrey County Council grant, for an annual season from Woking, and regular haulage behind steam every day was the enticing prospect before me. I shall always remember that first day when I joined the commuters on the 6.45am Salisbury, 8.47 off Woking with unrebuilt Merchant Navy 35004, which touched 88mph through Esher and passed Surbiton in eleven minutes flat, start to pass, for the twelve miles, a performance I never repeated in the following three years. And finding I had a choice for my evening train home and, on seeing 30457 *Sir Bedivere* on the buffer stops with the coaches of the 5.39 to Salisbury, I rejected its Standard 5, and waited for the Arthur to join the 6.09 eleven coach train to

Basingstoke. That was just the first of thirty-six runs I enjoyed behind that engine during my student years.

My studies would lead to a degree in German language and literature and involved a lot of reading of dramas, novels and poetry, for which many 'free' periods had been inserted into the weekly timetable, especially during the second and third years. Most students used them to adjourn to the college refectory to drink coffee rather than study in the library, and I often utilised the spare time for a run to Woking and back in the middle of the day – with an annual season ticket there was no further expenditure to be made. In the second and third years there was one day in which there were no lectures I had to attend and one free morning and another free afternoon so I would pile the books to be read into my brief case and go to Waterloo to spend a few hours going 'to and fro', getting runs behind as much steam as possible whilst reading one of the texts that had been set for my analysis. I did not time everything – after all, I did have some work to do – but I always took note of engine and load and starting time and if the run looked promising I would time it properly throughout.

During those three years, and a subsequent year (1960-1) of commuting to London when I worked as a clerk at Paddington before joining the Western Region management training scheme, I travelled behind all the Merchant Navies (414 runs), all bar one of the Bulleid Light Pacifics (890 runs), forty-eight of the fifty-four King Arthurs (585 runs), all the Lord Nelsons (383 runs), thirty-five of

Salisbury had two regular turns to Waterloo on semi-fasts, using one of its allocated 30448-30454 series on the 11.16am arrival and 2.54pm return and a Maunsell S15 on an afternoon up service and the 6.54pm down. Then in 1958, the S15s gave way to the King Arthurs on that also, and as withdrawals took place, 30796, 30798 and 30799 joined 30450, 30451 and 30453 in dominating those two diagrams. These trains were some of my favourites, always energetic with regular net times between 26 and 29 minutes for the 31 minute schedule, although we were often put to the slow line at Hampton Court Junction or Esher to allow 3pm and 7pm Waterloo–West of England trains to pass as they did not stop at Woking. You can tell how often I chose these trains for I had 55 runs behind 30453 *King Arthur* – more than any other steam locomotive – 44 runs behind 30451, 29 behind 30450, 39 behind 30798 and 28 behind 30796.

There was a Bournemouth–Waterloo 'stopper' that left Woking at 9.48am, ideal if my first lecture was at 11am, and this was booked for a Nine Elms King Arthur or Standard 5, although this was the train on which the remaining Urie H15s (30489 and 30491) were most likely to appear. Another favourite was the late night 11.15pm Waterloo–Basingstoke whose engine returned on a freight. It was booked for one of the four Nine Elms Scotch Arthurs, 30763, 30774, 30778 and 30779, but there was often a 'foreigner' and even a Bulleid pacific that was too rough for express work (it was on this train that the unrebuilt 35005

30453 *King Arthur* on the 2.54pm Waterloo – Salisbury at Waterloo, May 1960. (Author)

the forty Schools (323 runs), the twenty SR-allocated BR Standard 5s (399 runs), the four Basingstoke BR Standard 4 4-6-0s (88 runs), eighteen of the Maunsell S15s (51 runs), eleven of the twenty Urie S15s (18 runs), six of the taper-boilered H15s, 30473-478 & 30521-30524 (29 runs), all the Urie H15s, 30482-30491 (a further 10 making 63 in total including the earlier Surbiton runs). This included runs between Waterloo and Woking from 1961 to 1964 when I was travelling home regularly from my training on the Western Region and 'excursions' on Saturdays when I chose extra travel to London or to go to the Railway Enthusiasts' Club at Farnborough for the afternoon, returning back to Woking with a late evening semi-

fast that was regularly worked by a Urie S15 until the diagram changed and an Eastleigh BR Standard 2-6-0 took over. During that time, I have to say that loss of time by the locomotive due to poor steaming or a mechanical defect was rare. I cannot recall a single occasion in those 585 runs when a King Arthur let me down. I can remember one appalling run with an unrebuilt Merchant Navy just before its rebuilding, when clearly its mechanical condition was atrocious. A couple of Standards struggled for steam and I had a number of poor Lord Nelson runs on the 6.04am from Southampton Terminus which I used to catch from Woking for my early Thursday lecture – poor steaming again the problem.

30450 *Sir Kay* passes Woking Golf Course approaching Brookwood on the 2.54pm Waterloo–Salisbury, July 1960. (Author)

Six-wheeled tender 30805 *Sir Constantine* passes the Railway Enthusiast Clubroom at Farnborough with a relief Waterloo–Bournemouth express, July 1959, shortly after its transfer from the Kent Coast line after the latter's electrification. (Author)

nearly fell to bits). I used to travel home on this service after a table tennis league match (I was in UCL's second team) or the weekly 'MethSoc' student meetings at Hinde Street chapel near Bond Street, and always looked forward to fireworks on this four-coach train. I would get one of the 30521-30524 series often on this train too, and I remember an excellent run with 30521, although they were usually no match to the regular King Arthurs.

One night I was talking to the driver at Woking and he asked if I would be on the train later in the week – if so, as long as no Inspector was about (very unlikely at that time of night) I was welcome on the footplate. Of course I took him up on his offer and found Eastleigh's

30777 *Sir Lamiel* at the head of the 5.9pm Waterloo – Basingstoke commuter train in May 1960, which it hauled regularly from the autumn of 1959 until June 1960. The author had over fifty runs with this engine on this train. (Author)

30784 *Sir Nerovens* at the head of the usual four coaches. After we had cleared Vauxhall, the driver put me in his seat and I regret to say I made a bit of a mess of it. I shut off steam and braked for the 40mph Clapham curve and then tried to open the regulator in good time before we hit the gradient up to Earlsfield. The regulator was extremely stiff, however, and although I tugged with both hands, I needed the driver's extra muscle power and we had dropped to 25mph before we got *Sir Nerovens* with steam on again. We'd lost a couple of minutes by Wimbledon, but with so light a load and main line all the way to Woking, we were soon back to schedule. Although the engine was some time out of Shops, it rode well enough – the only problem was that stiff regulator – a problem I never found in all my official footplate work later on the Western Region. I had one other unofficial footplate trip on a King Arthur from Woking to Farnborough on a summer Saturday when I was on my way to the Railway Enthusiasts' clubroom there. The driver had obviously seen me around and at Woking as I was admiring 30777 *Sir Lamiel*: I was told to get on up and hide myself until we were clear of the platform. It was just a stopper to Salisbury with about six or seven on, but we proceeded easily enough with little rattle or shake and clearly the driver considered 777 to be an exceptionally strong engine. I did not volunteer to operate the regulator this time though.

My favourite train in the summer of 1959 and through to graduation in June 1960 was the 5.9pm Waterloo–Basingstoke, a ten-coach train that was normally hauled by one of Basingstoke's double-chimney Standard 4s, 75076-75079, in 1957 and 1958. At that time I usually chose to travel home on the 5 o'clock West of England train with its Exmouth Junction Merchant Navy and Salisbury top link crew. However, after the Kent Coast electrification and the Hastings dieselisation, a number of Eastern Section King Arthurs and Schools were transferred to Nine Elms and Basingstoke and in 1958-9 we were treated to Schools 30904, 30905, 30908, 30918 and 30923, and King Arthurs 30765, 30773, 30777, 30793, 30794 and 30795 on the 5.09 and its corresponding up commuter train in the morning, the 9.08am from Woking. From the autumn of 1959 right through to June 1960, 30777 *Sir Lamiel* became a fixture on this train and was superbly consistent, night after night getting its 375 gross ton train to Woking in 27-29 minutes net for the 24.4 miles, 31 minute schedule. I had 54 runs in all behind this locomotive on this train during that period. A selection of the best are included in the table (right), which also prints some good representative runs by other down semi-fast services with King Arthurs.

I've scoured my runs with the 5.09pm Waterloo and 30777 and find a remarkable consistency. I started opting for this train rather than the 5 o'clock from the autumn of 1959. During the first months of 1960 there was a severe pws to 15mph (reduced to 5mph for some weeks) during a bridge rebuilding at New Malden, and the Basingstoke drivers made it their aim to get to Woking in the

	30796- 72B *Sir Dodinas le Savage* 6 - 210t 2.54 W - Salis 8/10/59		30777- 70D *Sir Lamiel* 9 - 345t 5.09 W - B'stk 19/2/60		30453 - 72B *King Arthur* 6 - 210t 2.54 W – Salis 23/3/60		30777 - 70B *Sir Lamiel* 10 - 375t 5.09 W - B'stk 2/6/60		30777 - 70B *Sir Lamiel* 10 - 375t 5.09 W-B'stk 3/6/60	
Waterloo	0.00		0.00		0.00		0.00		0.00	
Vauxhall	3.08	48	3.30	52	3.16	58	3.26	53	3.29	56
Clapham Jcn	6.56	40*	7.08	40*	6.35	40*	6.59	42*	6.53	46*
Earlsfield	9.28	55	9.34	52	8.56	58	9.13	55	9.05	56
Wimbledon	11.17	64	11.33	62	10.45	25*sigs	11.06	64	10.59	61
New Malden	15.07	pw 5*	14.26	pw 10*	14.05	pw 15*	13.43	67	13.39	65
Surbiton	19.26	50	19.28	46	18.30	53	16.17	42* sigs	15.46	68
Hampton C Jcn	-	67	-	60	-	65	-	57	17.00	73
Esher	21.55	20* SL	22.02	71	21.12	15* SL	19.00	67	17.54	73
Hersham	-	51	-	73	-	50	-	68	-	73
Walton	25.23	58	24.25	72	24.07	58	21.30	70	20.10	75
Weybridge	27.29	63	26.14	71	26.17	63	23.28	67	21.58	72
W.Weybridge	-	74	-	78	-	73	-	77	-	70 eased
W. Byfleet	29.41	72	28.25	76	28.33	72	25.47	72	24.17	
Woking	32.29 (27 net)		32.04 (27½ net)		31.33 (27 net)		29.06 (27 net)		27.41	

	30765 - 70D *Sir Gareth* 8 - 255t 7.54 W - B'stk 30/6/60		30765 - 70D *Sir Gareth* 11 - 405t 5.09 W - B'stk 15/7/60		30796 - 72B *Sir Dodinas le Savage* 5 - 190t 6.54 W - Salis 7/2/61		30777 - 70D *Sir Lamiel* 4 - 150t 9.54pm W - B'stk 21/2/61	
Waterloo	0.00		0.00		0.00		0.00	
Vauxhall	3.42		3.10	50	3.01	58	3.11	57
Clapham Jcn	7.11	pw 10*	6.41	47*	6.11	46*	6.24	45*
Earlsfield	-	43	8.43	58	8.20	60	8.38	58
Wimbledon	13.32	53	10.35	62	10.37	15*sigs	10.36	63
New Malden	-	61	13.18	64	15.10	64	13.06	72
Surbiton	18.58	66	15.31	65	17.30	60*sigs	15.20	62* eased
Hampton C Jcn	-	72	-	70	18.50	71	16.42	64
Esher	-	73	17.48	70	20.02	43* SL	17.41	71
Hersham	-	75	-	70	-	61	-	76
Walton	23.21	76	20.14	70	23.00	66	20.01	78
Weybridge	25.11	72	22.10	67	24.58	70	21.41	80
W.Weybridge	-	81	-	73	-	76	-	82
W.Byfleet	27.20	74	24.31	70	27.13	73	23.49	73
Woking	30.53 (27½ net)		27.41		34.22 sigs (26 net)		27.07 (26 net)	

30765 *Sir Gareth*, 'star' of the run on the 5.9pm Waterloo – Basingstoke with eleven coaches on 15 July 1960 and one of the last survivors at Basingstoke, being withdrawn in September 1962. It is seen here at Southampton Central, just a month before withdrawal, 18 August 1962. (D. Clark)

scheduled thirty-one minutes, despite the enforced slowing while all nine coaches went over the bridge, which must have cost at least three to four minutes (I find the regular load was reduced to nine coaches, 345 tons gross during that period). On 8 February 1960, Driver Stephens and 30777 got to Woking in 32 minutes (28 net) with 66 on the level through Walton and 72 max in the dip at West Weybridge, on the next day with a different driver, 32 minutes again, with 65 at Walton and 73 after Weybridge; on 10 February, with yet another driver, 33 minutes including a signal check before the pws, with 64 at Walton and 71 after Weybridge; on the 15 February with Driver Ashton, 31½ minutes, with 70 at Walton, 74 at West Weybridge; on 16 February with Ashton again, 32 minutes with 69 at Walton and 73 at West Weybridge; on the 19th with Ashton once more, 32 minutes

(27½ net) after a very slow entry to Woking, with 73 at Walton and 78 at West Weybridge.

By May, with the New Malden bridge slowing still severe, the load was back up to ten coaches, 370 tons. Driver Ashton was again performing on 10 May when he again got to Woking in 32 minutes dead (28 net) with 66 at Walton, 73 at West Weybridge; on 18 May with Driver Hayward 31 minutes exactly (27 net) after he'd applied the brakes too hard and come to a dead stand at New Malden, 70 at Walton and 74 at West Weybridge; next day, with Hayward again, 31½ minutes, 69 at Walton and 72 West Weybridge, and on 23 May, with Driver Burden, 33½ minutes (29 net), with 68 at Walton and 71 at West Weybridge, after only 5mph at the New Malden bridge.

The bridge slowing was finally removed in June 1960 and another unknown Basingstoke driver on

2 June got 30777 and 370 tons to Woking in 29 minutes actual, although a net 27 minutes after a signal check to 45 at Surbiton (we'd caught up the Portsmouth electric that stopped at Surbiton). We'd recovered to 70 at Walton and 77 at West Weybridge. The next evening we were completely unchecked and completed the run in 27 minutes 41 seconds having reached a full 75mph at Walton, but with time so well in hand, we eased back and drifted at 70 past West Weybridge. On 7 June, 30777 had 11 coaches for 400 tons and another different driver who reached Woking in 29 minutes actual, with 68 at Walton and 72 at West Weybridge. Finally, on 8 June, the bar was raised higher again with 30777 through Surbiton in 15½ minutes at 65, 73 by Esher, dropping to 65 up the grade to Weybridge station and 73 in the dip afterwards, reaching Woking in 28 minutes 4 seconds.

Then, at the end of the month, new speed restrictions came on with a pws at Clapham Junction. On 7 July we had a Feltham S15, 30840, which took 32 minutes 29 seconds (29 net) with 66 at Walton and 73 at West Weybridge, a faster top speed than I've seen with an S15 published elsewhere (see table below), and on 15 July, 30765 *Sir Gareth* had 11 coaches for 405 tons and achieved 27 minutes 41 seconds with 70 sustained on the level from Hampton Court Junction to Walton and 73 in the dip after Weybridge (see log on page 225). I've also included another run with Salisbury S15 30831 and two H15 runs, the one with 30521 on the 11.15pm and one of my last runs with 30489 on the 1.54pm Waterloo–Basingstoke,

	30831 72B (Maunsell S15) 10 – 365t 5.09 W – B'stoke 4.5.60		30489 –70A (Urie H15) 5 – 180t 1.54 W – B'stoke 10/59		30521 –70A (Maunsell H15) 6 – 200t 11.15 W – B'stoke 10.5.60		30840 – 70B (Maunsell S15) 10 – 365t 5.9 W- B'stoke 6.7.60	
Waterloo	0.00		0.00		0.00		0.00	
Vauxhall	3.15		3.09		3.10		3.26	
Queens Road	-	51	-	53	-	58	-	50
Clapham Jcn	6.56	41*	6.28	40*	6.18	46*	7.17	pws 15*
Earlsfield	9.26	49	8.52	53	8.26	56	11.49	42
Wimbledon	11.30	56	10.58	59	10.15	66	14.08	56
New Malden	14.40	pws 15*	14.32	pws 15*	13.11	pws 15*	17.05	58
Surbiton	19.19	45	19.04	49	17.48	SL 25*	19.33	60
Hampton C Jcn	-	55	-	61	-	55	21.01	65
Esher	22.10	58	21.44	63	20.52	62	22.06	65
Hersham	-	60	-	63	-	68	-	66
Walton	25.00	60	24.28	62	23.23	70	24.43	65
Weybridge	27.17	57	26.44	59	25.23	66	26.46	64
W.Weybridge	-	68	-	66	-	67	-	73
W. Byfleet	29.50	66	29.21	61	27.49	68	29.07	72
Woking	33.22	(29 ½ net)	33.16	(29 net)	31.05	(27 net)	32.29	(29 net)

Urie H15 30489 arrives at Woking with the 9.48 to Waterloo, a stopping train from Bournemouth, June 1960. (Author)

The taper-boilered Urie H15, 30491, arrives at Woking with the 9.48am Woking–Waterloo, c1960. (Author)

which was normally a Standard 4 4-6-0. Unfortunately, I cannot find a log with a Salisbury S15 on the 6.54 Waterloo. My memory is that with a six-coach load, they were away smartly and would achieve around 62-64 mph before Hampton Court Junction, and around 55-60 on the slow line around Walton, with a top speed of around 65-70mph at West Weybridge.

I need to record one further episode on the 5.09 before moving on to performances in the up direction. Driver Carlisle at Basingstoke, a former GW man, knew of my interest and asked me one week in March 1959 if I was going to be on the 5.09 the following week as he was booked to it every day. He said that he was going to see how fast he could get to Woking and hoped he'd have different engines to test – a sort of unofficial trial. I managed three days and was lucky enough to get him with three different classes of locomotive that graced the train from time to time in 1959 – a Standard 4 4-6-0, a Schools and a King Arthur. All the runs were first class, with actual times varying from 25¾ to 26½ minutes and to my surprise 30794 *Sir Ector de Maris* was the winner by dint of an extraordinarily fast start and the maintenance of a steady 75mph after Surbiton. The logs of the three engines are given opposite.

The up trains most likely to get a Southern 2-cylinder 4-6-0 were the Woking departures at 9.08am (ex-Basingstoke), 9.48 (slow ex-Bournemouth) and the 10.42 (semi-fast from Salisbury). The first was booked for a Basingstoke Schools or King Arthur after the 1958-9 winter timetable (the balancing turn of the 5.09 Waterloo), the 9.48 was a Nine Elms King Arthur or Standard 5, and occasionally an H15, and the 10.42 was the Salisbury King Arthur that returned on the 2.54pm Waterloo. The table below has example logs from all these services plus one run on the afternoon up Salisbury semi-fast that balanced the 6.54 down and an exceptional run with 30788 on a Friday evening

	30794 – 70D *Sir Ector de Maris* 10/375t		75078 – 70D (Std 4 4-6-0) 10/375t		30923 – 70D *Bradfield* 10/375t	
Waterloo	00.00		00.00		00.00	
Vauxhall	02.51		03.12		03.24	
Queens Road	-	61	-	58	-	55
Clapham Jn	05.56	46*	06.20	45*	06.46	45*
Earlsfield	08.04	57	08.31	55	08.59	54
Wimbledon	09.52	66	10.24	64	10.52	65
New Malden	12.20	71	12.59	69	13.23	70
Surbiton	14.30	75	15.19	73	15.42	72
Hampton Court Jn	-	75	-	75	-	74
Esher	16.28	75	17.20	75	17.42	76
Hersham	-	76	-	77	-	76
Walton	18.48	76	19.35	78	19.58	77
Weybridge	20.31	73	21.12	75	21.36	74
West Weybridge	-	75	-	78	-	83
West Byfleet	22.46	73	23.22	75	23.33	81
Woking	25.45		26.18		26.22	

Driver Carlisle of Basingstoke and 30794 *Sir Ector de Maris* after achieving the winning time of 25¾ minutes from Waterloo, March 1959. (Author)

	30768 - 70A Sir Balin 9 - 260t 10.20arr ex S'ton 5/59		30799 - 72B Sir Ironside 7 - 250t 15.27arr ex Salis 12/12/59		30788 - 71A Sir Urre of the Mount 10 - 340t 18.04 arr(FO)ex S'ton 6/6/60		30457 - 70A Sir Bedivere 6 - 220t 14.04 arr ex B'stk 18/6/60		30840 – 70B (Maunsell S15) 9 - 340 9.39arr ex B'stk 2/5/60	
Woking	0.00		0.00		0.00		0.00		0.00	
W.Byfleet	4.02	66	3.56	68	4.02	63	4.03	65	4.25	60
Weybridge	6.14	69	6.10	70	6.29	65	6.28	63	6.49	65
Walton	-	73	7.56	78	8.21	77	8.30	71	8.47	70
Hersham	-	75	-	78	-	76	-	70	-	69
Esher	10.15	74	10.08	76	10.36	78	10.59	72	11.19	70
Hampton C Jcn	-	74	11.04	73	11.32	77	11.59	69	12.23	67
Surbiton	12.16	69	12.26	45* sigs	12.34	76	13.06	68	13.40	61
New Malden	14.20	71	15.38	60/pw 5*	14.39	70	15.16	67	17.56	pw 5 *
Wimbledon	16.39	67	20.04	35* sigs	16.59	67	17.38	69	22.42	53
Earlsfield	18.16	61	22.03	61	18.35	66	19.14	68	24.41	60
Clapham Jcn	3 min stand*sigs		24.09	20* sigs	20.22	40*	21.32	20* sigs	26.47	40*
Queens Road	26.41	42	-	52	-	35* sigs	-	46	-	51
Vauxhall	28.18		28.00		23.58	2 min sigs	26.00	3 mins sigs	30.29	sigs
Waterloo	31.09 (26½ net)		31.27 (26 net)		30.56 (26½ net)		34.22 (27½ net)		34.37 (28½ net)	

	30803 - 71A Sir Harry le Fise Lake 8 - 260t 10.20arr ex S'ton 1/7/60		30457 - 70A Sir Bedivere 9 - 265t 10.20arr ex S'ton 6/7/60		30795 - 70D Sir Dinadan 10 - 370t 9.39arr ex B'stk 5/4/61		30798 - 72B Sir Hectimere 7 - 250t 11.16arr ex Salis 26/3/62		30796 – 72B Sir Dodinas le Savage 7 – 250t 11.16arr ex Salis 22/1/60	
Woking	0.00		0.00		0.00		0.00		0.00	
W.Byfleet	4.26	62	4.18	66/70	4.16	68	3.49	69/73	4.09	64/66
Weybridge	7.03	60	6.40	69	6.44	66	6.05	72	6.41	61
Walton	9.03	74	8.32	75	8.42	74	7.50	78	8.42	70
Hersham	-	72	-	73	-	75	-	79	-	72
Esher	11.25	77	10.56	72	11.00	75	9.59	76	11.07	73
Hampton C Jcn	12.25	71	11.58	72	11.58	73	10.54	74	12.07	68
Surbiton	14.52	5* sigs	13.10	64*sigs	14.25	5* sigs	12.16	42* sigs	13.16	63
New Malden	-	50	15.28	67	18.10	55	16.25	pw 2*	16.20	pw 15*
Wimbledon	-	63	18.01	65	21.01	61	22.18	57	20.29	56
Earlsfield	-	65	19.44	68	22.41	67	24.02	64	22.12	67
Clapham Jcn	26.08	pw 10*	21.45	pw 25*	24.30	40*	25.50	43*	23.54	44*
Queens Rd	-	47	-	49	-	51	-	58	-	53
Vauxhall	30.35	15* sigs	25.53		28.21	15* sigs	28.51		27.27	sigs 10*
Waterloo	34.28 (28 net)		29.20 (27½ net)		32.05 (27½ net)		31.46 (25½ net)		31.14 (27½ net)	

Bournemouth relief train that probably excelled all the others. I didn't travel on the 9.08 very often because if I had a 10am start at College, I needed to catch the 8.47.

With an 11am start I could wait until the 9.48, so I had to make a special effort to catch the 9.08 and then hang around in London. It was ideal, however, if I needed to

cross Hungerford Bridge and buy some German texts at a bookshop close to Charing Cross station. I've tabled one run on the 9.08 with a Basingstoke King Arthur, 30795,

30779 *Sir Colgrevance* of Nine Elms arrives at Woking with the 9.48am to Waterloo, May 1959.
(Author)

and one with a Feltham S15 which very occasionally could be found on these turns – especially after 1962 when all the Maunsell passenger engines had been withdrawn. I had further runs with 30838, 30839 and 30840 on these commuter services as well as the Saturday 12.24 or 1.54 Waterloo–Basingstoke in 1963 and 1964. After 1962, however, the usual power was a BR Standard 5 as more were transferred from other Regions to the SR for their last years of steam in place of the Maunsell engines.

The speeds in the up direction were higher than on the down road, given the start from Woking and

30798 *Sir Hectimere*, transferred from Dover to Salisbury in June 1959, which became a 'regular' on the two Salisbury – Waterloo semi-fast diagrams between then and its withdrawal in June 1962, and behind which the author had thirty-nine runs, c1959.
(MLS Collection)

One of the last survivors, 30798 *Sir Hectimere*, departs Waterloo with the 6.54pm Waterloo–Salisbury semi-fast train, May 1962. 30798 was withdrawn at the end of July 1962. (ColourRail)

the gradual favourable gradient from Weybridge to Walton. The maximum speed could therefore be sustained on the level from Walton to Hampton Court Junction where trains tended to get checked or were already easing ready for the 60mph restriction from New Malden. As you see from the above, speeds in the 75-80mph area were not unusual with King Arthurs on the up journey – they were decidedly speedier than the Lord Nelsons on the 11-coach 7.51am Woking which rarely exceeded the mid-60s, mainly because of the initial impetus of a much faster start to West Byfleet.

The run with 30803 above reminds me of the informal tutorial I had at the end of finals in June 1960 with the professor who was one of the specialists in mediaeval German epic poetry that I'd been studying on my frequent rail trips. Many of the poems by writers like Hartmann von Aue and Walter von der Vogelweide were Arthurian legends inspired by the French Chrétien de Troyes and the British Geoffrey of Monmouth, and one in particular was about a Knight, Erec, who married a young woman and spent the next six months in bed with her, causing tongues to wag. When he heard the gossip he arose in a huff, donned his armour and slammed his visor down so hard that it jammed. His poor wife had to lead his horse as he sought adventures in the forest. Thereafter,

when they were challenged by other knights, his wife was required to point him in the right direction to see off his opponent. Inevitably in the end he suffered an injury and his wife had to tend him. She managed to prise his helmet open and he 'saw' – in other words, he came to his senses and returned back to Arthur's court, a chastened and now mature knight. This errant knight's full name was 'Erec fis du Lac', or when various vowel and consonant shifts between ancient German and English are deciphered, 'Harry le Fise Lake' (son of King Lake), our 30803 I'd travelled behind that very morning. The professor saw the aptness of this great steaming monster,

belching smoke and blinkered by its smoke deflectors charging through the countryside and exploded in mirth and our tutorial was concluded over a pint in the college bar.

Other characters in the legends that I was studying found themselves as names on the engines I was travelling behind. The French and German poets took the Celtic stories and gave them a bawdy sense of humour not unlike the style of Chaucer's *Canterbury Tales*. Poor *Sir Kay* (30450) was foster brother

of King Arthur and a grandfatherly figure who was somewhat portly and overweight, and who had to be lifted onto his horse with great difficulty and suffered the embarrassment at awkward moments of sliding backwards off the horse when it succumbed under his weight. *Morgan le Fay* (30750) was 'Morgan the Witch', the sister of Arthur who plotted with the villain, Sir Mordred, against Arthur – Sir Mordred incidentally nearly became 767 until someone tumbled to the fact that he was Arthur's

enemy. As I think is relatively well-known, *Elaine* (30747) and *Maid of Astolat* (30744) were one and the same person – and as the flighty young mistress of *Sir Launcelot* (30455) hardly the fitting object of one engine name, let alone two! Another pairing were *The Red Knight* (30755) and *Sir Ironside* (30799) who shared the same role. *Sir Dodinas le Savage* (30796) or in German *der wilde Dodinas* did not refer to his character, but only the fact that he did not belong to Arthur's Court – he was a foreigner,

30803 *Sir Harry le Fise Lake* was transferred to the Western Section in June 1959 and received a Urie bogie tender in November 1959, in which form it would have been the author's engine on the 6.04am Southampton Terminus before his college tutorial. (ColourRail)

Model Jidenco 'OO' kit-built model of T14, 30446, from the author's collection. (Author)

an 'outsider' who lived in the forest. And *Sir Uwaine* (30791), or *Sir Ivan* in German, owned a pet lion that used to horrify and offend the onlookers by assisting the knight in jousts with other knights, a most unfair tactic. However, when a group of knights set off for some adventure outside Arthur's castle, to rescue a damsel in distress or some other heroic deed, Sir Uwaine would be routinely asked, 'You will bring your lion with you, won't you?'

Basically the stories were about the impact of strangers bringing about change in Arthur's court, the latter representing any 'closed' society – school or church or business enterprise (or railway organisation?). Every legendary story seems to take place outside the confines of King Arthur's Round Table and castle. Inside the court, all activities are nothing but 'play' that change nothing but are merely practice for real deeds elsewhere. It is significant that in the stories King Arthur himself is but the figurehead, the organiser, who never takes part in the adventures himself. Apart from the boyhood stories with *Merlin* (30740) he only appears in the tragic story of the final battle when he is mortally wounded and he gives his sword *Excalibur* (30736) to be thrown into the lake by *Sir Bedivere* (30457). And I searched

Bottom to top, Wills kit 30796 *Sir Dodinas le Savage*, Hornby models 30803 *Sir Harry le Fise Lake* and 30453 *King Arthur*, Wills kits 30777 *Sir Lamiel* and 30755 *The Red Knight*, from the author's collection. (Author)

in vain for our survivor *Sir Lamiel*. The only mention in any of the stories is a passing reference to 'Sir Lamiel of Cardiff'. Perhaps 30777 should bear a Canton (86C) shedplate?

In the 1970s I found I needed to do something with my hands to give my brain a rest, and did something that as a boy I could never afford – I got myself a model railway layout and as well as buying both continental 'HO' and British outline 'OO' off the shelf, I bought second-hand models and detailed and repainted them in the form of locomotives I knew and had travelled behind. Before the days of the main manufacturers producing superb prototypes of almost every class, I also built kit models to cover some of the gaps. As well as Continental and Western Region collections, I have a good selection of Southern Region models detailed and painted in the era of the late 1950s when I was at college and travelling behind them. I have models of the following classes represented in this book:

30446 Drummond T14, Jidenco kit-built, painted BR black c1950, acquired 1982
30453 *King Arthur*, Hornby R2583, BR green, acquired 2007
30457 *Sir Bedivere*, Wills Finecast kit, large tender (ex-30490), BR green, acquired 1983
30755 *The Red Knight*, Wills Finecast kit, multi-jet chimney, BR green, acquired 1983
30764 *Sir Gawain*, Hornby R2581, BR green (weathered), acquired 2007
30777 *Sir Lamiel*, Wills Finecast kit, BR green, acquired 1982

30796 *Sir Dodinas le Savage*, Wills Finecast kit, Wills Q 6-wheel tender, BR green, acquired 1983
30803 *Sir Harry le Fise Lake*, Hornby R2582, 6-wheel tender, BR green, acquired 2007
30826, Maunsell S15, weathered BR Black, Hornby 23329 acquired 2016
30843, Maunsell S15, with 8-wheel flush-sided tender, BR black acquired 2016
32328 *Hackworth*, Nu-Cast kit, BR mixed traffic lined black, acquired 1980

I await the production of an H15. If I had known that decent King Arthur models had been coming on the market, perhaps I would have sought kits for this missing class instead, though, frankly, I think my air-brushed Brunswick

green is much closer to the original than Hornby's version.

David Charlesworth, Chairman of the Friends of the Darjeeling Himalayan Railway and railway artist, supported the Railway Children charity which I founded in 1995 and for which any royalties from this book will be donated, by allowing some of his paintings to be used as Railway Children Christmas cards. Some of the charity's trustees also commissioned paintings from him and let the charity use the images on their cards. I commissioned a painting of *Sir Lamiel* passing Surbiton where I trainspotted after school on the 5.9 Waterloo–Basingstoke on which I commuted so often, and the painting has been a popular Railway Children Christmas card for these last three or four years.

David Charlesworth's painting of *Sir Lamiel* passing Surbiton with the 5.9pm Waterloo–Basingstoke in March 1960, commissioned and owned by the author, and used as one of the images for the charity Railway Children Christmas cards in 2012-14.

Chapter 12

A FIREMAN'S ACCOUNT

David Solly was a fireman at Bricklayers Arms for nine years in the 1950s and 1960s. During that time had many turns firing King Arthurs, mainly the six-wheel tender variety which were most commonly used at that depot. He remembers in particular 30770, 30776-30778, and 30796 -30805. David is now seventy-seven, but recounts here a typical journey back around 1959.

'My driver and me book on with the time clerk, and move to the next window to see the list clerk to learn our booking-on time for tomorrow's duty. Then we go to the foreman's office to find out the number of our engine today. It's 30803 *Sir Harry le Fise Lake* on No.3 Old Shed, outside, over the pit. We have to prepare the engine. Having stowed our personal gear in the locker box, we go to the Stores, draw out thick and thin lubricating oil and paraffin for the headlamps, also an oil can with a long spout for reaching the big ends and side rods. I do the headlamps and gauge glass lamp, the rest is the driver's, for I have to test the water level in the boiler by pulling one of the levers attached to the two gauge glasses. Then I open the firedoors, deflect the shovel round at different angles – you can tell by the flames

what is properly burnt coal and what is not. I don't muck about, I get the pricker and rake through the firebox to get a proper foundation in the back corners and under the firedoor. I put coal down the sides and front of the firebox and gradually build up the fire between my other tasks. I put the blower on to force air through to liven up the fire quickly.

'I now leave the footplate to check the sandboxes, three on each side and two at the front of the tender for when running backwards with a train. Luckily ours are full, so I bring a handbrush, a 7/8in spanner and a shovel to remove any ash left over in the smokebox (I was checking that the "disposal" staff had done their job properly). I sweep around the ring of the smokebox door frame to ensure it is airtight, slam the door shut and tighten the nuts, then go back to see how my fire is doing. I rake through the fire again to shake any ash into the ashpan and shovel in another round of coal over the grate area, building up the back corners, close the firedoor and leave the blower on and open the damper a bit. I drop down into the pit and rake out the ashpan, then fill four headlamps, putting one at the rear of the tender and one on

the route destination disc over the left-hand buffer bracket. My driver then joins me and we move up to the water crane to top up. We test the whistle and move up to the "ringing-out" hut. The signalman is rung from there to advise him that we are light engine to Ewer Street (situated between London Bridge and Waterloo East), the other side of the flyover slope to Blackfriars. Ewer Street was a sub-depot of Bricklayers Arms with a turntable that could accommodate a "Schools" or other smaller 4-4-0 types. The depot consisted of a long island platform for continental fruit and vegetable vans, ten on each side, with road access to lorries.

'We chuff our way up to North Kent West Junction and swing past the wagon repair works and the twelve-road carriage shed, joining the main-line at North Kent East Junction Box. I now swap round the lamp and disc on our buffer beam bracket, checking carefully to see I don't touch the 700 volt live rail. I've been building up the fire en route to Ewer Street, keeping the boiler only half full of water, as I don't want the safety valves to blow off with two factories very close. Our load is all twenty of these continental vans with a British guard's van on the rear.

30803 *Sir Harry le Fise Lake* an Ashford-based locomotive, the locomotive in the run described by David Solly with the six-wheel tender seen in this photo. It is photographed here between 1948 and 1951 before it received the brunswick green livery. It was transferred to the Western Section in June 1959 and received a Urie bogie tender in November 1959. (J.M. Bentley Collection)

We pull the first ten vehicles out round the curve towards Cannon Street to two ground signals, then set back slowly to the other ten. At this moment we don't know what route we will be taking – we await Control's decision. We are restricted to 45mph, so they have to calculate the best path for us.

'We should be away at 12.15pm. I put on another round of coal, and fill the back corners, and fill the boiler, avoiding blowing off. The guard now advises us that we are routed via Swanley, Otford and Sevenoaks and Tonbridge to Dover Marine. I can now put up the appropriate route indicator discs – one halfway up on the smokebox bracket which says "spl" and has our duty number and one over the left bracket. The ground signal comes off, we give a short toot on the whistle to alert the guard and get the "right-away". I adjust the dampers for the ashpan at the front – about 1in open throughout the trip – and open the rear one fully, closing the firedoor to draw up the fire. We take it steadily through the points and crossovers past Metropolitan Junction and London Bridge station. I keep a watch on the train until we are clear of the platforms, as my mate has more than enough to do. Greens glow ahead and I fire again but leave the firedoors open with just a small gap. Before New Cross I rapidly throw a few shovelfuls over the front, but not too much. I watch the road ahead on my side till Lewisham flyover, then fill up the boiler and fire again. It's uphill to Hither Green at 1 in 250/1 in 140 all the way to Chislehurst Junction. I have fired twice more and the boiler water level is ¾ full – I don't want the safety valves to lift, as you can waste ten gallons of water.

'At Chislehurst Junction we

swing left on to St Mary Cray Junction and I fire another round. It is uphill to Swanley Junction at 1 in 100/132 for two miles and we take the Otford line, tackling the 1 in 100 to Eynsford, another half mile to the summit, then up and down hill until another summit at Shoreham. We are doing well now, pass Otford station and cross to the Sevenoaks branch, four miles to Bat and Ball station. At last I can sit and hold tight – our speed is about 20mph (it's a very twisty bit of track). If I tried to fire here I could easily be thrown across the cab and fracture a shoulder. At last we join the mainline to Tonbridge, so I turn the injector on to raise the water level to three-quarters. Time for another round on the fire. I close the firedoors to save us having a 'blow back' on entering the two mile long Sevenoaks Tunnel. I shut off the injector, and open the firedoor enough to light the gauge glass.

"I now assist my driver looking for the colour light repeater signal near the bottom of the tunnel wall and also the white lights from men working on the track. I operate the whistle, for the driver's view is very restricted by the tunnel wall. It is downhill now at 1 in 142/122 for over five miles. We rattle our way out, past Weald signalbox, keeping to the 45mph limit, I lean out as we round the slight curve for Hildenborough to check the vans behind us – all is still OK. We get the 'distant' on approaching Tonbridge and are stopped in the platform. We don't need to take water here, so I use the time to make up the fire which is torn about a bit. The driver is replenishing our tea cans. We are being held for an express from

Victoria to overtake us. Once we get away we climb for a mile and a half up 1 in 270/300 of Tudeley Bank, then down to Paddock Wood, where we again have to stop to let another train pass. Between Paddock Wood and Headcorn we have twenty changes of gradient! Even so, with a good crew and engine, it is a racing ground with passenger trains. I once reached 102mph here with West Country 34004 *Yeovil*.

'Off we go again, by Marden nearly five miles away, I'm busy firing again and replenishing the boiler. Our steam pressure has been no trouble. The safety valves on this engine blow off at 200 psi, I keep it at 180-200 which is ample for this turn. It's downhill to Staplehurst, the "distant" signals are off. We keep going until we get to Ashford and are routed into a platform to take water. (With no water troughs on the Southern and only a 4,000 gallon six-wheel tender, we need to fill up here.) I pull the coal forward in the tender, while we wait for the signal to clear. Once we get the "off" I start to fire again all round the grate – it's against the grade now for the next eight miles. I put the injector on as we climb up past Herringe signal box, it's just a further two miles to the summit at Westenhanger Racecourse station, then downhill all the way to our destination. I check to see if I need to fire again, close the firehole doors and at last sit down. However, I keep a beady eye on our speed – it would be easy to exceed our 45mph limit here. We're through Sandling Tunnel and onto the four-track section past Cheriton Halt, past Folkestone Junction, the small motive power depot and the

junction for the harbour branch. The fire is now burning down, and I fill the boiler for the last time – it helps if the fire is low for cleaning.

'We enter the Martello Tunnel and drift along the sea wall through the Abbotscliffe and Shakespeare Tunnels, and Dover engine shed. The line sweeps round almost a horseshoe towards Dover Priory but we keep straight on into the Marine Harbour. The shunter uncouples us and we pull down to the bottom of the harbour and back to the depot, where we are relieved by the "disposal" men. We are not booked to work anything back so we make our way to the Priory station and return home "passengers".'

It is clear from David Solly's description that even working a freight limited to 45mph was no sinecure – he seems to have been active 'multi-tasking' for much of the journey. David goes on to comment that although they had only three Arthurs based at Bricklayers Arms at that time – 30800-30802 – he regularly fired at least seven others and says that he never once was short of steam. The biggest load he had was a 120-wagon freight from Three Bridges to Norwood Junction. He says he could only just see the guard's green handsignal. He had to fire regularly, but not too much at any one time or it would take longer to burn through. He worried about the climb with that load from Redhill to Coulsdon up the 1 in 200 to Quarry Tunnel. From the glare from the firebox, he estimated their speed was down to 7 or 8mph – he says he could count the bricks in the tunnel!

Chapter 13

KING ARTHURS AND S15S IN PRESERVATION

King Arthur Class

30777 *Sir Lamiel*

777 Sir Lamiel was built at the North British Locomotive Works Glasgow, Works No. 23223 and entered service on the Southern Railway as E777 in June 1925. Like the other Scotch Arthurs it was allocated at various times to both the Southern's Eastern and Western Sections, but spent considerable periods at Nine Elms. It was one of three King Arthurs at Dover when the Kent Coast line was electrified and was then transferred to Feltham, and finally Basingstoke where it worked commuter trains to London and semi-fast trains to Waterloo and Salisbury. It was withdrawn in October 1961, having run 1,257,638 miles and stored at Eastleigh. 30453 *King Arthur* had been earmarked for the national collection, but when it was withdrawn the previous July, it was found to be in poor condition – its frames were said to be cracked – and there were no

777 *Sir Lamiel* at Nine Elms around the time of its record-breaking run from Salisbury to Waterloo on the Atlantic Coast Express, c1932. (J.M. Bentley Collection/F. Moore)

Carlisle–Hellifield Cumbrian Mountain Express, 22 5.82
11.50am Carlisle–Carnforth – 777 *Sir Lamiel* – 10 coaches, 375 / 395 tons

Location	Mins Secs	Speed	Schedule
Carlisle	00.00		T
Petteril Bridge Junction	04.01	15*	
Durran Hill South Sidings	-	pws 20*	
Scotby	-	35	
Cumwhinton	10.58	42	
Howes Siding	12.45	52	
Cotehill	-	47	
MP 300	-	46	
Armathwaite	18.38	64	
Baron Wood Tunnels	-	58/63	
Lazonby	24.25	55*	
Long Meg Sidings	26.33	54/58	
Little Salkeld	27.46	47	
Langwathby	29.31	48	
Culgaith	35.03	56/pws 20*	
Newbiggin	37.18	38/42 (blowing off)	
Long Marston	-	46/pws 30*	
		48/sigs 10*	
Appleby	48.12		6½ E
	00.00		12¾ L (Slow water)
Ormside Viaduct	-	53	
Ormside	04.20	49/43	
Helm Tunnel	-	40	
Griseburn	08.46	38	
Crosby Garratt	13.45	41/46	
Smardale Viaduct	-	36	
Kirkby Stephen	17.23	34½ /36	
Birkett Tunnel	-	34	
Mallerstang	-	37/33½ /27	
Ais Gill	30.53	32	6½ L
Shotlock Tunnel	-	45	
Garsdale	36.27		4 L
	00.00		6L
Dent	10.06	35 (priming)	
Dent Head	-	40	
Blea Moor Tunnel	19.29	32/35	
Ribblehead	-	pws 20*	
Selside	-	52	
Horton-in-Ribblesdale	30.02	58	
Helwith Bridge	-	64	
Stainforth Sidings	-	58*	
Settle	36.16	66	4¾ L
Settle Junction	38.38	62*/40*	
Long Preston	43.20	pws 20* (long)	
Hellifield	48.25		9½ L

30777 *Sir Lamiel* during its eight-year sojourn as a Dover engine, seen here c1952. (J.M.Bentley Collection)

Drummond 'watercart' tenders extant to allow it to be restored in original condition for the National Railway Museum. It was therefore scrapped in October 1961 and 30777 substituted for the national collection in its place. 777 held the record for its 1930s run on the *Atlantic Coast Express* and had the reputation as being one of the best

Preserved King Arthur 30777 *Sir Lamiel* in brunswick green and the final BR logo as it ran in the 1959-60 period when it regularly hauled the author's commuting trains between Waterloo and Woking, here at the Great Central Railway Autumn Gala at Loughborough, October 2013. (Author)

Overhauled and repainted in SR Bulleid malachite green, 777 *Sir Lamiel* in pouring rain at the Great Central Steam Gala, October 2014. (Gordon Heddon)

A close up view of the preserved 777's cab in malachite green at the GCR Gala, October 2014. (Gordon Heddon)

of the Scotchmen.

The engine was placed in the care of the 5305 Locomotive Association in 1978 and restored to main line operating condition. It operated railtours on the Settle and Carlisle and on the Leeds–York-Scarborough route in the 1980s and on one occasion worked a York–Leeds scheduled local passenger train when the DMU failed. It has sported three liveries – SR olive green, BR brunswick green and, since its overhaul at Tyseley in 2012, SR malachite green. It is

based at the Great Central Railway at Loughborough and is currently available for main line operation.

This locomotive featured in the author's travels between Waterloo and Woking in 1959-1961, as recounted in Chapter 11, and on one railtour on the Settle and Carlisle *Cumbrian Mountain* Express in May 1982, the log of which is printed on page 240.

Urie S15 class

30499

30499 was built at Eastleigh in May 1920 and was withdrawn

from Feltham shed on January 1964, having run 1,241,024 miles. It resided at Woodham Brothers Barry until bought by the Urie Locomotive Society and stored on the Watercress Line. It is currently dismantled at Ropley pending restoration, with a target date of completion of 2020.

30506

30506 was built at Eastleigh in October 1920 and was withdrawn from Feltham shed in January 1964, having run 1,227,897 miles. It was sold to Woodham Brothers in July and arrived at their Barry scrapyard

An early shot of Urie S15, 499, with stovepipe chimney and Southern Railway green livery c1933, after receiving smoke deflectors but before exchanging its bogie tender for a 'watercart' from a withdrawn Drummond 4-4-0. (J.M.Bentley Collection)

in October 1964. It was sold to the Urie Locomotive Society in April 1976 and restored and put into service in dark green as Southern Railway 506 in July 1987. After fourteen years' operation on the Mid Hants line, it was withdrawn from operations in 2001 and is being overhauled, ready to return to service in 2015.

Maunsell S15 class

30825

30825 was built at Eastleigh in April 1927 and was withdrawn from Salisbury shed in January 1964, having run 1,384,665 miles. It was sold to Woodham Brothers and purchased by the Essex Locomotive Society. In 1981, its boiler was sold to the Mid Hants Railway for use on the restoration of 30506. Its frames and wheels went to Shipyard Services in Essex and when the North Yorkshire Moors Railway S15 30841 was found to have defective frames,

its boiler was transferred to 30825 to enable its restoration and use in place of the other engine. 825 was then restored with a Maunsell 8-wheel flush-sided tender, and ran on the North Yorkshire Moors Railway in green SR livery until its

boiler ticket expired in June 2013. It is currently being overhauled at Grosmont.

30828 *Harry A. Frith*

30828 was built at Eastleigh in July 1927 and was withdrawn from Salisbury shed (where it had spent its whole existence) in January 1964, having run 1,287,124 miles. It

30506 at its home Feltham shed shortly before withdrawal in January 1964.
(John Scott-Morgan Collection/A. Gosling)

Maunsell S15 830
ex works in Southern
Railway dark green livery,
between an L12 and an
Adams 0395 0-6-0 goods
engine, on Eastleigh shed,
March 1935.
(J.M. Bentley Collection)

resided in the Woodhams Brothers
scrapyard at Barry until March 1981,
when it was bought with the help of
a grant from the Eastleigh Borough
Council by the Eastleigh Railway
Preservation Society for £10,500 –
exactly the same as its building cost
in 1927! It was restored in 1994 and
has worked on the East Somerset
and Mid Hants lines and was named
in 1996 after Harry Frith who had
long been associated with its rescue.
It then worked on the Swanage
Railway until 2002 and moved to

Ropley on the Watercress Line in
2004 for overhaul, which is planned
for completion around 2018.

30830

30830 was built at Eastleigh in
August 1927 and was withdrawn
from Feltham in July 1964, having
run 1,259,236 miles. It had operated
from Salisbury depot from 1927
until December 1963. It was
sold to Woodham Brothers and
moved to their Barry scrapyard in

December 1964. It went initially to
the Bluebell Railway in 1987 and
was sold to the Essex Locomotive
Society for restoration on the
North Yorkshire Moors Railway in
2000. It is currently stored without
a boiler at Grosmont awaiting
restoration.

30841 *Greene King*

30841 was built at Eastleigh in
July 1936 and was withdrawn
from Feltham in January 1964,

Manningtree to March, 3.4.76
7.04am Liverpool Street – Loughborough
841 *Greene King*
9 coaches 310/345 tons

Location	Mins Secs	Speed	
Manningtree	00.00		CE/ 37.266 off
Summit 1 in 134/189	-	32	
Bentley	-	40/31/53	
Ipswich	<u>17.28</u>		
	00.00		
Sproughton Box	03.51	48	
Bramford	-	51	
Claydon	07.30	47/52	
Needham Market	11.54	52	
Stowmarket	16.20	50	
Haughley Junction	20.03	20*	
Elmswell	27.15	30/53	
Thurston	32.41	58/50/54	
Bury St Edmunds	39.35	pws 15*/10*/32	
Saxham & Risby	46.25	42/54	
Higham	50.53	58	
Kennett	54.00	65	
Chippenham Junction	57.43	25*	
Snailwell Junction	59.18	48/63	
Fordham	62.23	68	
Soham	65.50	10* SLW	
Sutton Bridge Junction	-	52	
Dock Junction	73.23	20*	
Ely	74.25	30*/46/40*	
Chettisham	79.16	57	
Black Bank	81.03	60	
Manea	86.38	60/50 easy	
Stonea	89.08	48/53	
March	<u>96.11</u>		CE/ 31.133/151 on

having run 837,002 miles. From 1948 until September 1963 it had been based at Exmouth Junction. It was in comparatively good condition and in 1972 went to the East Anglian Transport Museum in Chappel. It was restored at Chappel and Wakes Colne and went into service on the Stour Valley Railway in SR green livery and named *Greene King* in 1974. It ran in the Shildon Rail 150 Cavalcade in 1975, going to the Nene Valley Railway in 1977 and the North Yorkshire Moors Railway in 1978. It was steamed there until its heavy overhaul was due and, after stripping down, it was discovered that its frames were out of true, and its boiler and other fittings were transferred to assist the restoration of 30825. The remains of 30841 now lie derelict at Grosmont, with no plans for overhaul at the present time.

The author joined a railtour from Liverpool Street to the Great Central Railway at Loughborough in April 1976, and 841 took over from a Class 37 diesel-electric at Manningtree and worked the tour train to March where a couple of class 31 diesels took over. The steam portion of the tour's log is printed left:

Maunsell S15 30841, c1950.
(J.M.Bentley Collection)

30847

30847 was built at Eastleigh in December 1936, the last engine of the class. After stationing at Exmouth Junction, Salisbury and Redhill, it was withdrawn from Feltham in January 1964, having run 931,829 miles. It was purchased from Woodham Brothers Barry for £9,720 in September 1978 jointly by the Maunsell Locomotive Society

and the '847 Group' and went to the Bluebell Railway. Now owned by the Maunsell Locomotive Society, it was restored in SR green livery with an 8-wheel flush-sided Maunsell tender and returned to service there in December 2013. It is currently operational.

Preserved and restored in the Southern Railway 1936 green livery in which it first appeared, and named *Greene King* after the Brewery sponsors, Maunsell S15 841. (John Scott-Morgan Collection/Photomatic)

Preserved S15 847 in Southern Railway livery at Horsted Keynes station, on the 3.02pm train to East Grinstead, during the Bluebell Model Railway Weekend, 28 June 2015 (Author)

Chapter 14

CONCLUSIONS

Urie transformed the London & South Western Locomotive fleet, giving it a sturdy and reliable stud of engines capable of withstanding the First World War intensive use and lack of maintenance and haulage of the increased wartime loads. Whilst the Drummond 4-4-0s continued work within their excellent capabilities for many years, his 4-6-0s were over-complex and inadequate in many ways and Urie worked quickly to satisfy the urgent needs of his company. Urie's engines were strong and heavy and remained in traffic, with little modification, for over forty years, and after initial teething problems with draughting and steaming, and front-line work in the 1920s, performed reliably on freight, mixed traffic and secondary passenger duties until the late 1950s (the N15s) and the early 1960s (H & S15s).

Maunsell took Urie's basic 2-cylinder 4-6-0 designs and met again the immediate needs of the Southern Railway in its difficult early days, by applying the principles he'd learned from Churchward and at Ashford and as Director of the ARLE. His 2-cylinder 4-6-0s, like those of Urie, were straightforward uncomplicated engines, robust and reliable, if not particularly sophisticated or advanced enough for Bulleid's inventive mind and taste, but he left them well alone and they carried on with sterling work throughout the Second World War and nationalisation with an availability that underpinned the brilliant but mechanically more 'fragile' Bulleid Pacifics.

Maunsell sought more novel new designs in the late 1920s, but his 4-cylinder Lord Nelsons although designed for 500 ton continental boat trains at a 55mph average speed and on paper significantly more powerful, never exceeded the best work of the King Arthurs, which were rostered to do anything that a Nelson could operate. With the lack of investment opportunities on the Southern Railway in the 1930s, due to the Company's electrification policy and the impact of the Depression, no further Nelsons were deemed necessary, and the King Arthurs continued as the mainstay of the Railway's express motive power, shared with the Schools for specific Kent Coast and Hastings line duties, until the Bulleid Pacifics were available in large numbers.

Maunsell did plan some outline proposals for more advanced locomotives that might have overtaken the roles of some of his 2-cylinder 4-6-0 designs – a Beyer-Garratt design was contemplated for the heavy Kent Coast Continental expresses, a 4-8-0 heavy freight engine was considered for haulage of the Kent coal traffic and a 2-6-2 design for mixed traffic work got as far as an outline drawing – but none were pursued in the context of the financial situation and the adequacy of Maunsell's 2-cylinder engines. Outline drawings of these three designs are included in the Appendix.

In the mid-1950s the Southern Region of British Railways received twenty BR Standard Class 5 4-6-0s and these took over some of the secondary passenger work from the N15s, H15s and King Arthurs. As 2-cylinder simple 4-6-0s with outside Walschaerts valve gear, they had their ancestry in Urie's 1914 H15s, the first engines of that general design layout in Britain. 73080-73089 and 73110-73119 were never loved by enthusiasts in the same way that the old Arthurs retained the affection of many, even though the SR publicity people tried to popularise them by resurrecting the names of the twenty Urie N15s, probably the most attractive names from the Arthurian legends. The Southern Region engines, unlike some of their Western Region peers, were never painted in the passenger

lined green livery, but retained the BR mixed traffic black – usually pretty filthy – until the end.

In looking back at the performance of the Urie and Maunsell engines, especially the King Arthurs, there is no doubt that their peak was during the 1930s, hauling the Southern Railway's accelerated services to Exeter and Bournemouth. The articles by Cecil J. Allen in *Railway Magazine* of the 1930s are full most months of logs of King Arthurs as well as LNER Pacifics and GW Kings and Castles. Obviously management would monitor carefully the punctuality performance of their key trains, but the cost of these was another major area of essential interest to the Boards of the 'Big Four' during times of financial stringency. Cecil J. Allen published some interesting cost comparisons in his British Locomotive Performance and Practice article in the June 1939 *Railway Magazine* which does not show the Southern Railway engines in too good a light. Perhaps it was because the Southern Railway had maintained a large number of older types of non-standard locomotives whereas the GWR in particular and the LMS had been standardising their fleets for a number of years.

The published costs and other economic data were:

Annual coal consumption costs, 1938 (average cost per locomotive mile)

GWR	6.65d
LNER	7.64d
LMSR	7.70d
SR	8.78d

Average cost per locomotive

GWR	£507

It was a 50/50 chance whether the 9.48am Woking–Waterloo (a stopper from Bournemouth) would appear behind a Nine Elms Standard 5, as here with 73118 *King Leodegrance* or a Maunsell King Arthur or H15 4-6-0, c1959. (Author)

LNER	£581	
LMSR	£639	
SR	£675	

Working expenses including wages (average cost per locomotive mile)

GWR	19.89d
LMSR	20.76d
LNER	21.23d
SR	21.82d

Average cost per locomotive

GWR	£1,517
LNER	£1,614
SR	£1,677
LMSR	£1,724

Maintenance, repairs and renewals (average cost per locomotive mile)

LMSR	3.72d
GWR	3.74d
SR	3.95d
LNER	4.61d

Average miles per locomotive per year

LMSR	28,879
GWR	26,852
LNER	25,740
SR	25,714

It is a pity that most of the Urie and Maunsell passenger and mixed traffic engines were withdrawn before the end of 1962 and at that time most were cut up at Eastleigh, Ashford and Brighton Works. It was only around 1964 that the Southern Region sought to sell their redundant engines to scrap dealers and the few S15s that went to Woodham Brothers in Barry were fortunate in being the only ones to survive, though luckily the authorities had identified a King Arthur as being worthy of preservation as part of the national collection along with a Schools and Lord Nelson. There are many new enthusiast proposals to construct brand new locomotives to cover significant classes that became extinct. It seems a shame and unwarranted that no-one has proposed to build a Urie H15 'Chonker', whose design was at the forefront of its type and is therefore of more historical significance than many of the other engines that we are so fortunate to have still among us. But there we are, I still have my memories of 30489 rumbling me to college on the 9.48 up from Woking and I can look at 777 in all its present day glory and reflect when it backed on to the 5.09 at Waterloo and I would think, 'Oh, it's 30777 again, that's my fifty-third run', as I erased my notation against that engine in my Ian Allan ABC threatening to wear right through the paper after so many previous rubbings out…

APPENDICES – WEIGHT DIAGRAMS, DIMENSIONS AND STATISTICS

1. Urie H15 – Weight Diagram and Dimensions

Statistics

Engine No.	Built	Withdrawn	Engine No.	Rebuilt	Withdrawn
30482	2/14	5/59	30330	10/24	5/57
30483	3/14	6/57	30331	11/24	3/61
30484	4/14	5/59	30332	11/24	11/56
30485	6/14	4/55	30333	12/24	10/58
30486	12/13	7/59	30334	1/25	6/58
30487	1/14	11/57	30335	11/14	6/59
30488	3/14	4/59			
30489	5/14	1/61			
30490	6/14	6/55			
30491	7/14	2/61 * Tapered boiler			

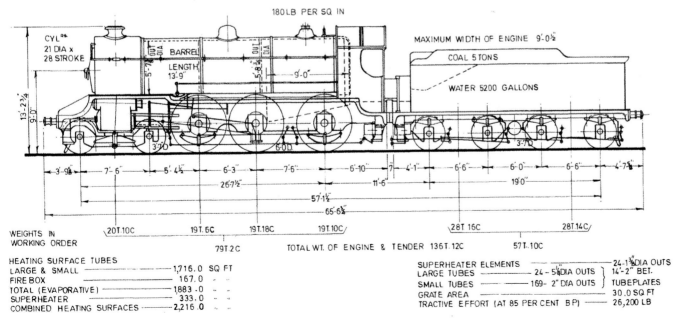

180LB PER SQ IN

CYL⁺ˢ 21 DIA x 28 STROKE

BARREL LENGTH 13'-9"

MAXIMUM WIDTH OF ENGINE 9'-0½

COAL 5 TONS

WATER 5200 GALLONS

9'-0"

13'-2¾" 9'-0"

3'-7 D 6'-0 D 3'-7 D

3'-9½" — 7'-6" — 5'-4½" — 6'-3" — 7'-6" — 6'-10" — 7' 4'-1" — 6'-6" — 6'-0" — 6'-6" — 4'-7¼"

26'-7½" 11'-6" 19'-0"

57'-1½"

65'-6¾"

WEIGHTS IN WORKING ORDER

20T.10C 19T.6C 19T.18C 19T.10C 28T.16C 28T.14C

79T.2C TOTAL WT. OF ENGINE & TENDER 136T.12C 57T.10C

HEATING SURFACE TUBES		
LARGE & SMALL	1,716.0	SQ FT
FIRE BOX	167.0	,, ,,
TOTAL (EVAPORATIVE)	1,883.0	,, ,,
SUPERHEATER	333.0	,, ,,
COMBINED HEATING SURFACES	2,216.0	,, ,,

SUPERHEATER ELEMENTS	24-1⅜"DIA OUTS	
LARGE TUBES	24- 5¼DIA OUTS	14'-2" BET.
SMALL TUBES	169- 2" DIA OUTS	TUBEPLATES
GRATE AREA	30.0 SQ FT	
TRACTIVE EFFORT (AT 85 PER CENT B P)	26,200 LB	

2. Urie H15 built by Maunsell with tapered boiler – Weight Diagram and Dimensions

Statistics

Engine No.	Built	Withdrawn	Engine No.	Built	Withdrawn
30473	2/24	8/59	30521	7/24	12/61
30474	2/24	4/60	30522	7/24	9/61
30475	3/24	12/61	30523	9/24	12/61
30476	4/24	12/61	30524	9/24	2/61
30477	5/24	7/59			
30478	6/24	3/59			

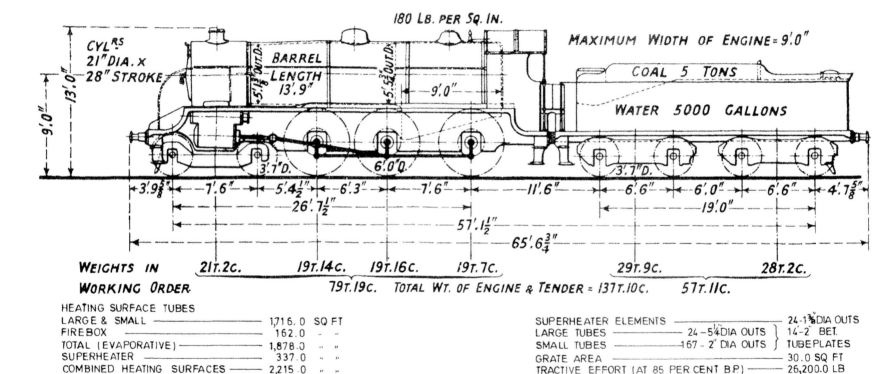

HEATING SURFACE TUBES
LARGE & SMALL — 1,716.0 SQ FT
FIREBOX — 162.0 " "
TOTAL (EVAPORATIVE) — 1,878.0 " "
SUPERHEATER — 337.0 " "
COMBINED HEATING SURFACES — 2,215.0 " "

SUPERHEATER ELEMENTS — 24-1⅜ DIA OUTS
LARGE TUBES — 24-5¼ DIA OUTS } 14'-2" BET.
SMALL TUBES — 167-2" DIA OUTS } TUBE PLATES
GRATE AREA — 30.0 SQ FT
TRACTIVE EFFORT (AT 85 PER CENT B.P.) — 26,200.0 LB

3. Urie N15 – Weight Diagram and Dimensions

Statistics
For names, see Chapter 5, page 41

Engine No.	Built	Withdrawn	Engine No.	Built	Withdrawn
30736	8/18	11/56	30746	6/22	10/55
30737	10/18	6/56	30747	7/22	10/56
30738	12/18	3/58	30748	8/22	9/57
30739	2/19	5/57	30749	9/22	6/57
30740	3/19	12/55	30750	10/22	7/57
30741	5/19	2/56	30751	11/22	6/57
30742	6/19	2/57	30752	12/22	12/55
30743	7/19	10/55	30753	1/23	3/57
30744	9/19	1/56	30754	2/23	1/53
30745	11/19	2/56	30755	3/23	5/57

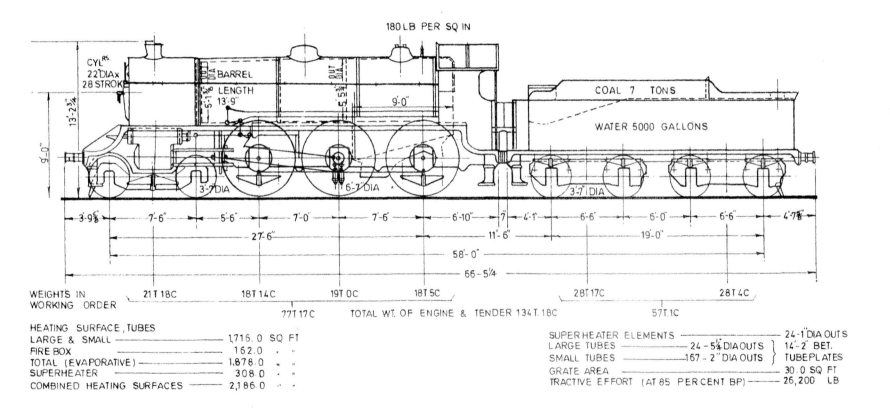

HEATING SURFACE, TUBES

LARGE & SMALL —— 1,716.0 SQ FT
FIRE BOX —— 162.0 " "
TOTAL (EVAPORATIVE) —— 1,878.0 " "
SUPERHEATER —— 308.0 " "
COMBINED HEATING SURFACES —— 2,186.0 " "

SUPERHEATER ELEMENTS —— 24-1" DIA OUTS
LARGE TUBES —— 24-5¼" DIA OUTS } 14'-2" BET.
SMALL TUBES —— 167-2" DIA OUTS } TUBE PLATES
GRATE AREA —— 30.0 SQ FT
TRACTIVE EFFORT (AT 85 PER CENT BP) —— 26,200 LB

4. Maunsell N15 King Arthur – Weight Diagram and Dimensions

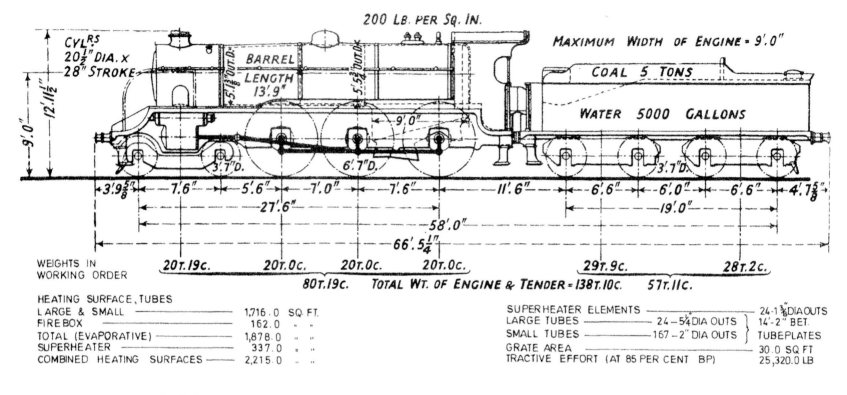

HEATING SURFACE, TUBES		
LARGE & SMALL	1,716.0	SQ. FT.
FIREBOX	162.0	" "
TOTAL (EVAPORATIVE)	1,878.0	" "
SUPERHEATER	337.0	" "
COMBINED HEATING SURFACES	2,215.0	" "

SUPERHEATER ELEMENTS	24–1⅜″ DIA OUTS
LARGE TUBES	24–5¼″ DIA OUTS
SMALL TUBES	167–2″ DIA OUTS
GRATE AREA	30.0 SQ FT
TRACTIVE EFFORT (AT 85 PER CENT BP)	25,320.0 LB

Statistics

For names, see Chapter 8.1, pages 63, 66-67 & 69

Eastleigh Arthurs, 'rebuilt' from Drummond G14 and P14

Engine No.	Built	Withdrawn	Engine No.	Built	Withdrawn
30448	5/25	8/60	30453	2/25	7/61
30449	6/25	12/59	30454	3/25	11/58
30450	6/25	9/60	30455	3/25	4/59
30451	6/25	6/62	30456	4/25	5/60
30452	7/25	8/59	30457	4/25	5/61

Scotch Arthurs, built by North British Locomotive Company, Glasgow

Engine No.	Built	Withdrawn	Engine No.	Built	Withdrawn
30763	5/25	10/60	30778	6/25	5/59
30764	5/25	7/61	30779	7/25	7/59
30765	5/25	9/62	30780	7/25	7/59
30766	5/25	12/58	30781	7/25	5/62
30767	6/25	6/59	30782	7/25	9/62
30768	6/25	10/61	30783	8/25	2/61
30769	6/25	2/60	30784	8/25	10/59
30770	6/25	11/62	30785	8/25	10/59
30771	6/25	3/61	30786	8/25	8/59

30772	6/25	9/61	30787	9/25	2/59
30773	6/25	2/62	30788	9/25	2/62
30774	6/25	1/60	30789	9/25	12/59
30775	6/25	2/60	30790	9/25	10/61
30776	6/25	1/59	30791	9/25	5/60
30777	6/25	10/61*	30792	10/25	2/59

* Preserved

Six-wheeled tender Arthurs built Eastleigh – Weight Diagram and Dimensions

Engine No.	Built	Withdrawn	Engine No.	Built	Withdrawn
30793	3/26	8/62	30800	9/26	8/61
30794	3/26	8/60	30801	10/26	4/59
30795	4/26	7/62	30802	10/26	7/61
30796	5/26	2/62	30803	11/26	8/61
30797	6/26	6/59	30804	12/26	2/62
30798	6/26	6/62	30805	1/27	11/59
30799	7/26	2/61	30806	1/27	4/61

5. N15X – Weight Diagram and Dimensions

Statistics

Engine No.	Built	Rebuilt	SR No.	BR No.	Withdrawn
327	1914	4/35	2327	32327	1/56
328	1914	2/36	2328	32328	5/55
329	1921	12/34	2329	32329	7/56
330	1921	9/35	2330	32330	8/55
331	1921	4/36	2331	32331	7/57
332	1922	11/35	2332	32332	1/56
333	1922	6/35	2333	32333	4/56

For names see page 95

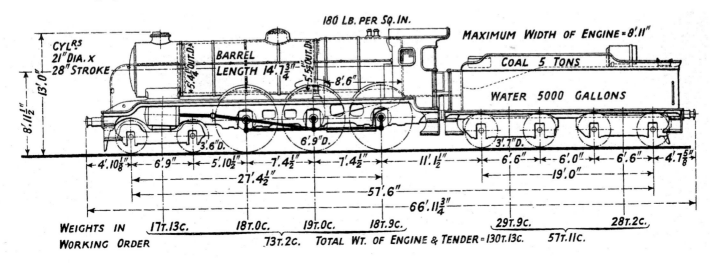

6. Urie S15 – Weight Diagram and Dimensions

Statistics

Engine No.	Built	Withdrawn	Engine No.	Built	Withdrawn
30496	5/21	6/63	30506*	10/20	1/64
30497	3/20	7/63	30507	11/20	12/63
30498	4/20	6/63	30508	12/20	11/63
30499*	5/20	1/64	30509	12/20	7/63
30500	5/20	6/63	30510	1/21	6/63
30501	6/20	6/63	30511	1/21	7/63
30502	7/20	11/62	30512	2/21	3/64
30503	8/20	6/63	30513	3/21	3/63
30504	9/20	11/62	30514	3/21	3/63
30505	10/20	11/62	30515	4/21	7/63

* Preserved

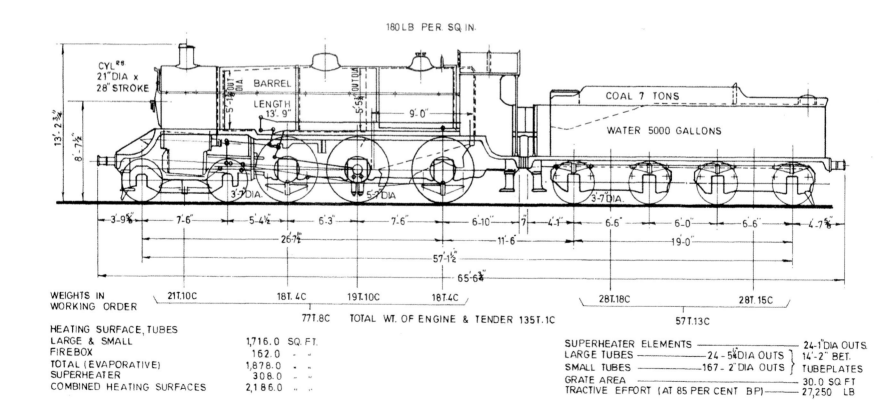

7. Maunsell S15 – Weight Diagram and Dimensions

Statistics

Engine No.	Built	Withdrawn	Engine No.	Built	Withdrawn
30823	3/27	11/64	30836	12/27	6/64
30824	3/27	9/65	30837	1/28	9/65
30825*	4/27	1/64	30838	5/36	9/65
30826	5/27	12/62	30839	5/36	9/65
30827	6/27	1/64	30840	6/36	9/64
30828*	7/27	1/64	30841*	7/36	1/64
30829	7/27	11/63	30842	8/36	9/65
30830*	8/27	7/64	30843	10/36	9/64
30831	9/27	11/63	30844	10/36	6/64
30832	10/27	1/64	30845	10/36	7/63
30833	11/27	5/65	30846	11/36	2/63
30834	11/27	11/64	30847*	12/36	1/64
30835	12/27	11/64			

*Preserved

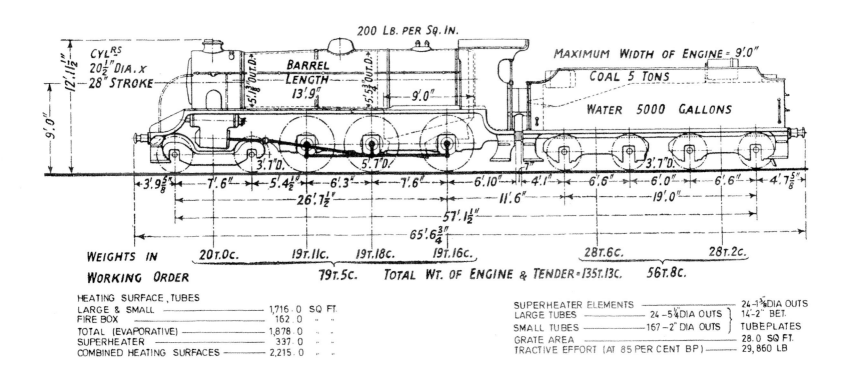

8. Drawings and/or Dimensions of the Proposed Maunsell Designs that were not constructed

4-6-2 + 2-6-4 6-cylinder Beyer Garratt mixed traffic locomotive

No drawing or weight diagram of the proposed Beyer Garratt is available

Proposed Dimensions

Cylinders (6)	16ft x 26in	Heating Surface	2,700sqft
Bogie wheel diameter	3ft 1in	Superheater	700sqft
Coupled wheel diameter	6ft 3in	Grate area	51.6sqft
Length over buffers	100ft 0in	Boiler pressure	220 lb
Height	13ft 1in	Weight	209tons 10cwt
Boiler diameter	6ft 9in	Coal capacity	7 tons
Boiler length	13ft 8¾in	Water capacity	6,000 gallons

4-8-0 4-cylinder heavy freight locomotive

Proposed Dimensions

Cylinders (4)	16½in x 26in	Grate area	33sqft
Bogie wheel diameter	3ft 1in	Boiler pressure	200lb
Coupled wheel diameter	5ft 1in	Weight	144tons 4cwt
Boiler diameter	5ft 6¼in–5ft 9in	Length over buffers	70ft 0¾in
Boiler length	13ft 9in	Height	13ft 1in
Heating Surface	1,989sqft	Coal capacity	5 tons
Superheater	376sqft	Water capacity	5,000 gallons

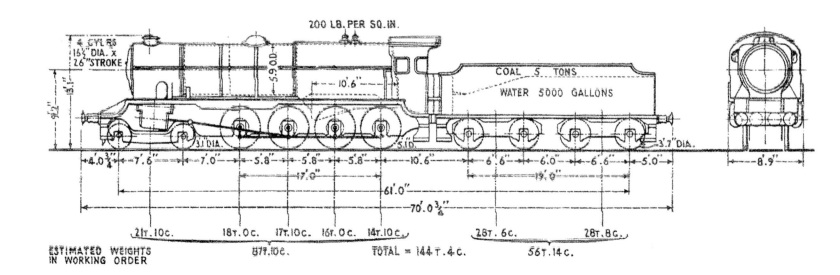

2-6-2 3-cylinder mixed traffic locomotive

Proposed Dimensions

Cylinders (3)	18in x 28in	Heating Surface	2,465sqft
Leading truck wheel	3ft 1in	Superheater	400sqft
Coupled wheel diameter	6ft 3in	Grate area	40sqft
Trailing truck wheel	3ft 7in	Boiler pressure	220 lb
Boiler diameter	5ft 7½in–5ft 9in	Weight	148 tons 14cwt
Boiler length	18ft 0in	Coal capacity	5 tons
Wide firebox length	9ft 0in	Water capacity	5,000 gallons

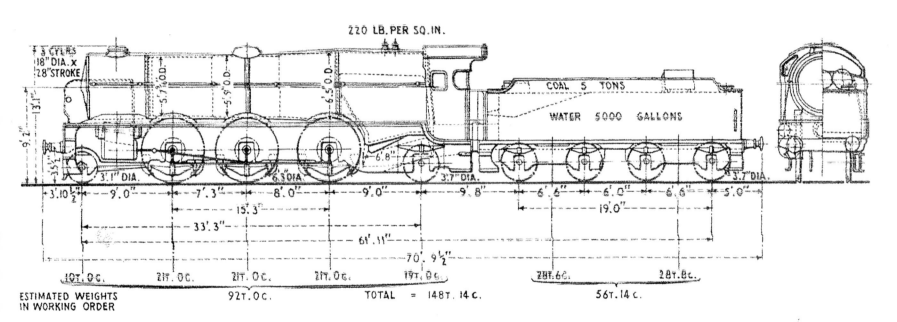

BIBLIOGRAPHY

Allen Cecil J., 'British Locomotive Practice and Performance'- Railway Magazine 1929-1943

Bradley D.L., Illustrated History of LSWR Locomotives – the Urie Classes, Wild Swan, 1987

Bradley D.L., Locomotives of the London and South Western Railway, RCTS, 1965

Chacksfield, J.E., Richard Maunsell, an Engineering Biography, Oakwood Press, 1998

Derry, Richard, The Book of the King Arthur 4-6-0s, Irwell Press, 2008

Esau, Mike, Steam into Wessex, Ian Allan, 1971

Fairclough, Tony & Wills, Alan, Southern Steam Locomotive Survey, the Urie Classes, D.Bradford Barton, 1977

Haresnape, Brian, Maunsell Locomotives, Ian Allan, 1977

Nock, O.S., The Southern King Arthur Family, David & Charles, 1976

Rogers, Colonel H.C., Steam from Waterloo, David & Charles, 1985

Steam Days, Feltham Shed & Marshalling Yard, Andrew Wilson, Redgauntlet Publications, March 2007

Steam Days, Urie and Maunsell class S15 4-6-0s, Roger Fredericks, Redgauntlet Publications, November 2010

Swift, Peter, Locomotives in Detail 4, Maunsell 4-6-0 King Arthur Class, Ian Allan, 2005

Swift, Peter, The Book of the H15 and S15 4-6-0s, Irwell Press, 2012

Townroe, S.C., Arthurs, Nelsons & Schools at Work, Ian Allan, 1973

Townroe, S.C., Famous locomotive types No.4 'King Arthurs and Lord Nelsons', Ian Allan, 1949

Trains Illustrated, Salisbury – Exeter, Cecil J. Allen, Ian Allan, November 1950

Trains Illustrated, From Switzerland to the Southern, Cecil J. Allen, Ian Allan, September 1956

Trains Illustrated, The Pacific & Baltic Tanks of the LB&SCR, R.C. Riley, Ian Allan, January 1957,

Whiteley, John & Morrison, Gavin, The Power of the Arthurs, Nelsons & Schools, Oxford Publishing Company, 1984

Back Cover Photos:

King Arthur 30794 Sir Ector de Maris reaches the summit of the 1 in 100 gradient between Beckenham and Shortlands Junctions, passing Downs Bridge with a Victoria-Ramsgate express, 26 August 1957. The photo was taken from a 'hole in the fence' of the photographer's garden which led onto the cutting, previously allotments.
(Ken Wightman)

King Arthur 30452 Sir Meliagrance leaving Exeter Central with a stopping train for Salisbury, 29 June 1957.
(R.C.Riley)

INDEX